Ready For 7th Grade

Math

A Summer Learning Activity Workbook For Students Entering Grade 7 [Ages 12-13]

Written and edited by Matthew Bowling

Printed in the United States of America

This paper meets the requirements of ANSI/NISO Z39.48-1992 (Permanence of Paper).

Paperback
ISBN 13: 978-1-5167-3156-5
ISBN 10: 1-5167-3156-5

TABLE OF CONTENTS

Introduction

Congratulations on finishing 6th grade! That is a big accomplishment, and you should be proud of yourself.

As you get ready to move into **7th grade**, it is normal to feel a little nervous. This workbook is designed to help you feel prepared and confident as you take on new challenges. You've got what it takes to succeed.

This workbook is organized into chapters that review important 6th grade skills and introduce new concepts you'll see in 7th grade. Each chapter starts with a short lesson to refresh your memory or even introduce new things before you start the daily activities. You can always go back to the lesson page if you need extra help.

The workbook is made to be completed over 10 weeks, with each day taking no more than 15 minutes. That is just enough time to practice without feeling overwhelmed. As you work through the chapters, the activities will gradually become more challenging, helping you strengthen your thinking skills and prepare for the work you'll do in 7th grade.

Sometimes you may work with an adult, and other times you may work independently. Either way, you're building skills, confidence, and strong learning habits. Get ready, because 7th grade is going to be a great year!

Chapter 1: Numbers and Operations

Natural numbers include all positive integers. These are often referred to as counting numbers since this is how you naturally count. **Whole numbers** include the positive integers and zero. 0, 1, 2, 3... and so on, are all whole numbers. **Integers** are whole numbers and their opposites. Some examples include 7, 0, and −56. **Rational numbers** can be positive or negative and include fractions, improper fractions, terminating decimals, and some repeating decimals. If it can be written as a fraction, it is rational. Some examples include 6, −23, $\frac{1}{4}$, and 0.46. Numbers are considered **irrational** when they cannot be expressed as a ratio of two integers. Essentially, numbers that cannot be expressed as fractions are irrational. For example, π is irrational because when it is expressed as a decimal it never terminates. Decimals that never terminate cannot be expressed as fractions, therefore pi is irrational.

Real numbers can fall into subsets such as natural numbers, whole numbers, integers, and rational numbers. Real numbers that are not rational are considered to be irrational numbers. Irrational numbers cannot be expressed as the ratio of two integers. In other words, irrational numbers cannot be expressed as a fraction.

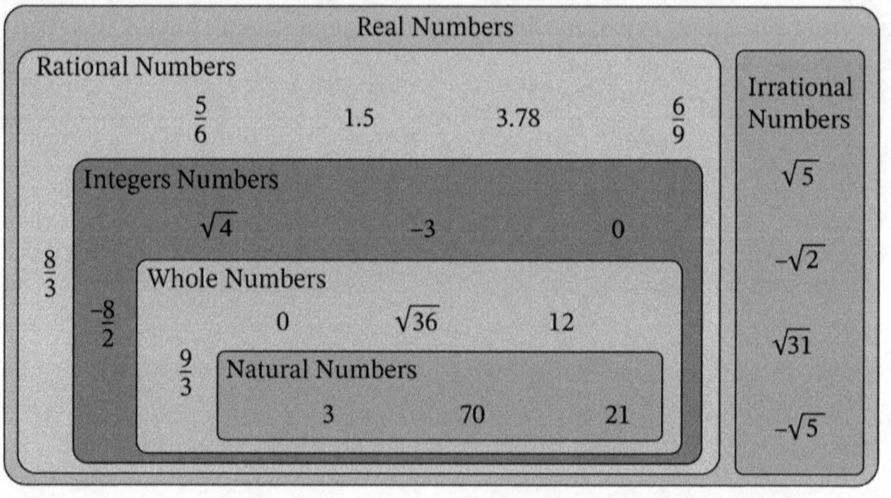

When ordering a set of real numbers, it is important to start by **converting** all the values to the **same form**. In general, it is most convenient to convert all numbers into decimal form rounded to the same place value.

It is often necessary to **estimate** the value of irrational numbers. For example, the value of square roots of numbers that are not perfect squares can be achieved using estimation. When estimating the value of $\sqrt{5}$, think about a number line with all the perfect squares listed in order. Then think about which two numbers it falls between. On a number line, $\sqrt{5}$ falls between $\sqrt{4}$ and $\sqrt{9}$. This means that $\sqrt{5}$ fits between the numbers 2 and 3 ($\sqrt{4}$ is 2 and $\sqrt{9}$ is 3). This means that $\sqrt{5}$ is between 2 and 3 on a number line, but closer to the number 2.

It is important to be able to add, subtract, multiply, and divide rational numbers quickly and easily. This includes operations with whole numbers, integers, and rational numbers, including fractions and decimals. The **order of operations** is a mathematical term to describe the correct sequence of steps to simplify an expression or solve an equation. The order of operations is very specific and important to remember. The acronym **PEMDAS** is used to help remember the order of operations.

The first operation is **parentheses**. If there are parentheses within other parentheses you should simplify from the inside out. The next operation is **exponents**. Again, if there are exponents inside other exponents, you should simplify from the inside out. The next group is **multiplication** and **division**. If multiplication and division are present, simplify from left to right. The last group is **addition** and **subtraction.** When there is more than one addition or subtraction operation, they are solved or simplified from left to right.

An **exponent** is used to identify a multiple of the same factor. The symbol for an exponent is a small number to the top right of a whole number. The small number shows how many times the whole number should be multiplied by itself. For example, 2^3 is in exponent form and is equivalent to $2 \times 2 \times 2$ which is expanded form. It is also equivalent to 8 which is standard form. When identifying equivalent expressions using exponents, you may be required to simplify exponents, expand exponents, or generate exponents.

A **root**, such as a square root, is another way of writing a fractional exponent. Instead of using a superscript, roots use the radical symbol ($\sqrt{\ }$) to indicate the operation. Nearly all of the roots you encounter will be square roots. A square root is the same as a number raised to the one-half power. When we say that a is the square root of b ($a = \sqrt{b}$), we mean that a multiplied by itself equals b: ($a \times a = b$). A **perfect square** is a number that has an integer for its square root. There are 10 perfect squares from 1 to 100: 1, 4, 9, 16, 25, 36, 49, 64, 81, 100 (the squares of integers 1 through 10).

A number's **opposite** is the number that is opposite to it on a number line or has an opposite sign. The opposite of -53 is 53. The opposite of 187 is −187. They both have the same distance from zero on a number line, just in different directions. The **absolute value** of a number is its distance from zero. Absolute value is notated by placing a vertical line on each side of a number. For instance, $|-46|$ means the absolute value of -46. Numbers and their opposite have the same distance from zero. -2 and 2 have a distance of 2 from zero. So, the absolute value of -2 is 2, or $|-2| = 2$. Also, the absolute value of 2 is 2, or $|2| = 2$. This may be noted by words, or with symbols.

The number 4,000 is in **standard decimal form**. In scientific notation this value becomes 4.0×10^3. The benefit of **scientific notation** is that extremely large or extremely small decimal values can be expressed in a concise form. In scientific notation the decimal needs to go to the right of the first non-zero number. To determine the power of 10, start at the original decimal point and count how many spaces it takes to arrive at the new decimal location. It is important to note that powers of 10 can be negative. For example, if the given number is smaller than 1, and the decimal point moves to the right, then the power of 10 will be negative. When a number is in scientific notation, the power of 10 will be used to determine how many places to move the decimal point, and in which direction.

Complete one worksheet each day in this chapter. Each day should take no longer than 15 minutes to complete.

Day 1

Classifying Numbers

Look at the numbers in the table below. Put each number in the correct classification.

7	-4	0	$\frac{3}{5}$	-9.2
$\sqrt{16}$	π	$0.3\overline{3}$	-15	$\sqrt{5}$

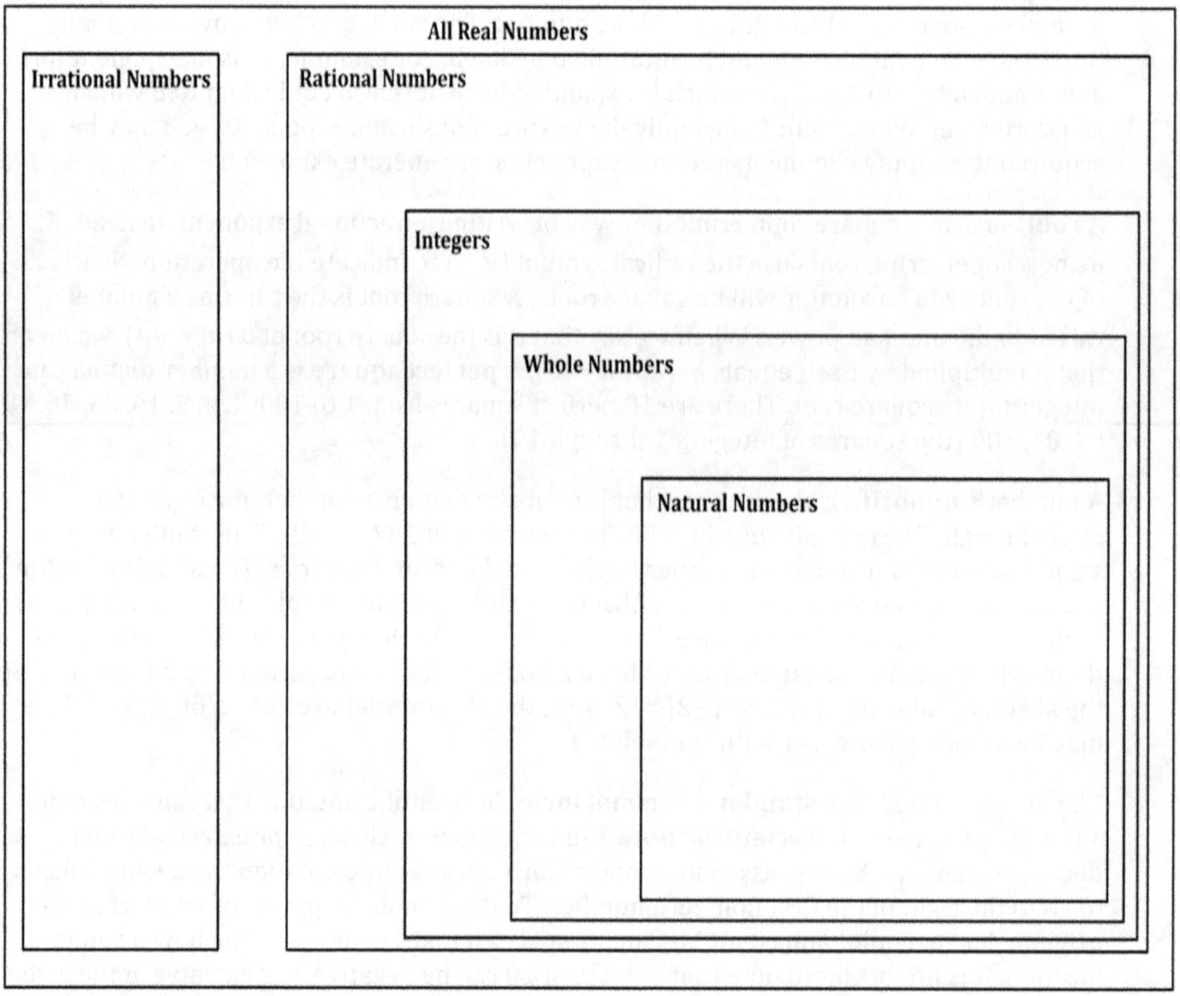

Day 2

Absolute Value and Opposites

Fine the absolute value of each.

1. $|-9|$

____

2. $|12|$

____

3. $|-3.5|$

____

4. $|\frac{2}{3}|$

____

5. $|0|$

____

Find the opposite of each.

6. 8

____

7. -4

____

8. 0

____

9. $-\frac{4}{7}$

____

10. 5.6

____

Day 3

Exponents

Evaluate the following problems.

1. 3^2

2. 5^3

3. 10^2

4. 2^4

5. $(-4)^2$

6. $(-2)^3$

7. -3^5

8. -4^2

9. $6^2 \div 3$

10. $2^3 \times 2^2$

Day 4

Square Roots

Find the square root of each problem below.

1. $\sqrt{49}$

____

2. $\sqrt{100}$

____

3. $\sqrt{9}$

____

4. $\sqrt{64}$

____

5. $\sqrt{36}$

____

6. $\sqrt{1}$

____

7. $\sqrt{81}$

____

8. $\sqrt{4}$

____

9. $\sqrt{16}$

____

10. $\sqrt{25}$

____

Day 5

Scientific Notation

Write the following numbers in scientific notation.

1. 45,000

____

2. 6,700,000

____

3. 0.0032

____

4. 0.00089

____

5. 9,010

____

Write the following numbers in standard form.

6. 3.2×10^4

____

7. 4.5×10^6

____

8. 7.1×10^{-3}

____

9. 9×10^{-2}

____

10. 1.09×10^3

____

Day 6

Basic Operations

Solve the following problems.

1. $-6 + 9$

2. $-12 - 5$

3. $8 - (-3)$

4. $-4(6)$

5. $-\frac{30}{5}$

6. $\frac{3}{4} + \frac{1}{8}$

7. $\frac{5}{6} - \frac{1}{3}$

8. $3(-\frac{2}{5})$

9. $4.2 + (-7.8)$

10. $\frac{-15}{-3}$

DAY 7

APPROXIMATING VALUES

Determine which two integers each irrational number below falls between.

1. $\sqrt{18}$

_________ and _________

2. π

_________ and _________

3. $\sqrt{56}$

_________ and _________

4. $\sqrt{72}$

_________ and _________

5. $\sqrt{90}$

_________ and _________

Approximate the value of each irrational number below. Round to the nearest thousandths place.

6. $\sqrt{50}$

____

7. $\sqrt{10}$

____

8. $\sqrt{20}$

____

9. $\sqrt{150}$

____

10. $\sqrt{66}$

____

Day 8

Ordering Real Numbers

Order the numbers from least to greatest.

1. $-3, 0, 5, -8, 2$

2. $4.2, 4.02, 4.22, 4.002$

3. $-\frac{1}{2}, -0.6, -\frac{3}{4}, -0.25$

4. $\sqrt{9}, 2.9, \pi, 3.1$

5. $-\sqrt{16}, -3, -4.1$

DAY 9

COMPARING REAL NUMBERS

Compare the values using the symbols >, <, or =.

1. -4 ☐ -9
2. $\frac{3}{4}$ ☐ 0.8
3. $\sqrt{16}$ ☐ 4
4. -2.5 ☐ -2.05
5. π ☐ 3.14
6. $-\sqrt{36}$ ☐ -6
7. 0 ☐ -0.1
8. -7 ☐ $|-7|$
9. 12^2 ☐ 24
10. 63 ☐ 7^9
11. $\frac{2}{7}$ ☐ π
12. -34 ☐ $-(2^2)$

Day 10

Order of Operations

Solve the following problems using order of operations.

1. $(5 + 3 \times 4 - 2) \div 5$

____

2. $(8 - 3)^2 + 6$

____

3. $\frac{12 \div 3 + 2}{3}$

____

4. $(184 + 56) \div [2(8 \times 3)]$

____

5. $5(4 + 2^2)$

____

6. $3^2 + 4^2$

____

7. $18 - (6 \div 2)^2$

____

8. $10 - 2(3 + 1)$

____

Chapter 2: Expressions, Equations, and Inequalities

A **variable** is a letter used to represent an unknown numerical value in an expression or an equation.

An **expression** includes numbers, variables, or numbers and variables separated with operation signs. Anything that stands alone or could belong on one side of an equal sign, is an expression. $3x + 4$ is an example of an expression. It contains numbers, a variable, and an operation. A single number is also considered an expression. An **equation** will always contain an equal sign.

To simplify an expression, combine like terms and follow order of operations. **Like terms** are terms that have the same letter and same power. If the terms do not have a variable at all, they are like terms as well. If they match, you can combine them. For example, $3x + 2x$ simplifies to $5x$, and $6 + 2$ simplifies to 8. However, you cannot combine $3x + 6$, so it is already in simplest form. Furthermore, you cannot combine $5y + 3x$ since the variable letters are different. It is also in simplest form.

Given the equation $-6 + x = 10$, we can solve for x by performing operations to isolate x and get our variable on one side of the equation by itself.

In this example, we want to get x by itself on one side of the equation. We can do this by adding 6 to both sides of our equation.

$$\begin{aligned} -6 + x &= 10 \\ +6 &= +6 \\ x &= 16 \end{aligned}$$

We can do this because if we add, subtract, multiply, or divide the same value on both sides of the equation, the sentence will still be true since it is balanced. For instance, if $2 = 2$, then $2 + 3 = 2 + 3$.

$$\begin{aligned} 2 &= 2 \\ 2 + 3 &= 2 + 3 \end{aligned}$$

Now we know that the solution for $-6 + x = 10$ is $x = 16$. This means that if x equals 16, the sentence will be true.

The last strategy would be to use the values of x to test the solution. In this example, we have 16 as a possible solution, so we can use 16 in place of x to see if the equation is true.

$$\begin{aligned} -6 + x &= 10 \\ -6 + (16) &= 10 \\ 10 &= 10 \end{aligned}$$

When we simplify with 16 as the x-value, we get 10 on both sides of our equal sign, so our sentence is true and 16 is the solution.

An **inequality** is a math sentence that connects two expressions with a comparison. These expressions are compared using a **greater than**, >, **less than**, <, **greater than or equal to**, ≥, or **less than or equal to**, ≤, sign.

There are some situations that can be illustrated by inequalities. In these scenarios, we need to identify what we know, what we don't know, and what the comparison is. Inequalities can be solved in a way similar to solving equations, using inverse operations.

Inequalities can be represented on a number line. Inequalities are two expressions connected by a greater than, less than, greater than or equal to, or less than or equal to sign. We express solutions to these sentences using number lines and a series of arrows, open circles, and closed circles. Having an "or equal to" sign indicates that a closed circle is needed. If the sign is simply greater than or less than, an open circle is used. The shaded part of the line represents the numbers that are included in possible answers for the variable. When you multiply or divide by a negative number on both sides of the inequality, the inequality sign must flip directions.

When there is a constant relationship between units, the rate can be described algebraically using the equation $y = kx$, where k represents the **constant of proportionality,** or the value of the relationship between x and y.

A **linear relationship** is a relationship between x and y values where the relationship is constant. Linear relationships are expressed using the equation $y = mx + b$, where m represents the **slope** of the line and b represents the ***y*-intercept**, or the point where the line passes the y-axis.

The slope of a line refers to its steepness. In other words, **slope** is the vertical change compared to the horizontal change, and it is often referred to as $\frac{\textbf{rise}}{\textbf{run}}$. This relationship can be described using the formula $m = \frac{y_2 - y_1}{x_2 - x_1}$, where m represents the slope, and x and y represent ordered pairs on the line.

Linear relationships can be described verbally or using tables, graphs, and equations.

A proportional relationship involves two quantities that vary directly. In an equation, this follows the form $y = kx$, where x and y vary directly based on the value of k. For example, the equation $y = 0.5x$ represents a proportional relationship. Every value of y is directly related to the corresponding value of x. Values for y are calculated by multiplying the corresponding values of x by 0.5.

In contrast, non-proportional relationships follow the form $y = mx + b$, where m represents the slope, and b is not equal to zero. The value of y is determined by multiplying or dividing the corresponding x-value by a consistent amount, but then a constant will be added or subtracted (the value of b). The difference between proportional, $y = kx$, and non-proportional relationships, $y = mx + b$, in equation form is that the variable b is not 0 in non-proportional relationships.

Complete one worksheet each day in this chapter. Each day should take no longer than 15 minutes to complete.

Day 11

Writing Expressions

Write out an expression numerically for each situation.

1. Sarah has some stickers. She buys 8 more stickers. Let *s* represent the number of stickers she started with.

____

2. A number is multiplied by 6. Let *n* represent the number.

____

3. Tyler had some money. He spent $12. Let *m* represent the amount he started with.

____

4. Emma earns $5 per hour. Let *h* represent the number of hours she works.

____

5. Three times a number, increased by 2. Let *t* represent the number.

____

6. A movie ticket costs $10 plus a $3 service fee. Let *t* represent the number of tickets.

____

7. A gym charges $25 to join and $15 per month. Let *m* represent the number of months.

____

8. A number squared, then increased by 4. Let *x* represent the number.

____

Day 12

Simplifying Expressions

Simplify the following expressions.

1. $7y + 3x + 5x - 2y$

2. $4a + 6 - 2a$

3. $9m - 3 + 7(4m)$

4. $5x + 3 + 2x - 1$

5. $2(12b - 7b) + 4$

6. $6n + 8 - 2n + 5$

7. $15 - 3x + 6(2x)$

8. $10p - 4 + 3 - 2p$

9. $7x + 2y + 3x$

10. $4(2x + 3)$

DAY 13

WRITING ONE-VARIABLE EQUATIONS

Write out an equation numerically for each situation.

1. A gym charges a $30 sign-up fee and $25 per month. The total cost that Angie spent was $155. Let *m* represent the number of months Angie paid for.

2. Liam had some money. After receiving $18 for his birthday, he now has $52. Let *m* represent the amount of money he had before.

3. A librarian had already placed 120 books on the shelves before her morning shift began. During the shift, she added 35 more book each hour. By the end of the shift, there were 245 books on the shelves. Let *h* represent the number of hours she worked during the shift.

4. A babysitter charges $15 per hour and earned a total of $105. Let *h* represent the number of hours worked.

5. James bought a water bottle. After using a $7 coupon, the final price was $15. Let *p* represent the original price of the water bottle.

6. Kerry buys several notebooks that cost $6 each and one planner for $12. She spent $54 in total. Let *n* represent the number of notebooks she bought.

7. A runner had already completed 1 miles before joining her team for practice. She then ran at a steady pace of 6 miles per hour. When practice ended she had run a total of 13 miles. Let *h* represent the number of hours she ran during practice.

8. Sofia had some books, but she gave 9 books to a friend. She now has 21 books left. Let *b* represent the number of books she originally had.

Day 14

Solving One-Variable Equations

Solve the following equations.

1.

$$x + 7 = 12$$

2.

$$x - 9 = 5$$

3.

$$3x = 18$$

4.

$$x \div 5 = 6$$

5.

$$2x + 4 = 14$$

6.

$$5x - 10 = 20$$

7.

$$4x + 3 = 19$$

8.

$$6x - 7 = 11$$

9.

$$\frac{x}{3} + 2 = 6$$

10.

$$4(x - 2) = 16$$

Day 15

Writing One-Variable Inequalities

Write out an inequality numerically for each situation.

1. A movie theater only allows children under 12 years old to attend a special matinee. Let x represent a child's age.

2. A recipe calls for at least 3 cups of flour. Let f represent the number of cups of flour used.

3. A hiker plans to walk for several days. Each day, she walks 4 miles, and she wants to walk at least 28 miles total. Let d represent the number of days she hikes.

4. A coffee shop charges $4 per latter plus a $2 service fee for each order. The total cost must be less than $30. Let n represent the number of lattes ordered.

5. A bus can hold no more than 50 passengers. Let p represent the number of passengers on the bus.

6. A video game requires players to have more than 500 points to reach the next level. Let s represent a player's points.

7. A personal trainer recommends 3 sets of push-ups for each client. Each set must include at least 15 push-ups, and a client wants to complete more than 60 push-ups. Let p represent the umber of push-ups per set.

8. A concert organizer bought 480 tickets to sell. Tickets are sold in packs of the same size, and each pack can contain no more than 60 tickets. Let t represent the number of tickets per pack.

Day 16

Solving One-Variable Inequalities

Solve the following inequalities.

1. $x + 5 > 12$

2. $x - 4 \leq 9$

3. $3x < 18$

4. $\frac{x}{2} \geq 7$

5. $2x + 3 > 11$

6. $5x - 10 \leq 15$

7. $4x + 1 \geq 13$

8. $6x - 7 < 11$

9. $-3x > 12$

10. $\frac{x}{16} \geq -2$

DAY 17

GRAPHING ONE-VARIABLE INEQUALITIES

Solve and graph the following inequalities on the number lines below.

1.

$$x - 3 \leq 1$$

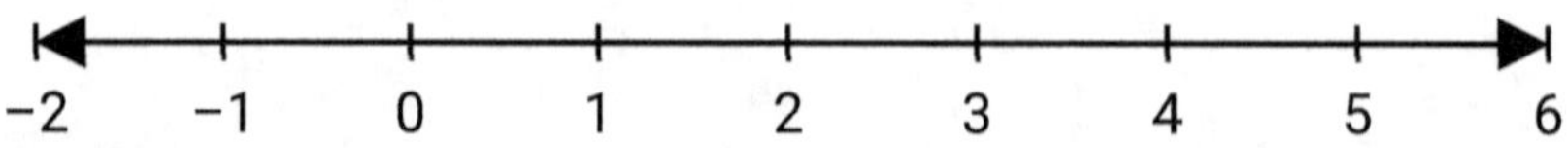

2.

$$2x - 6 > 12$$

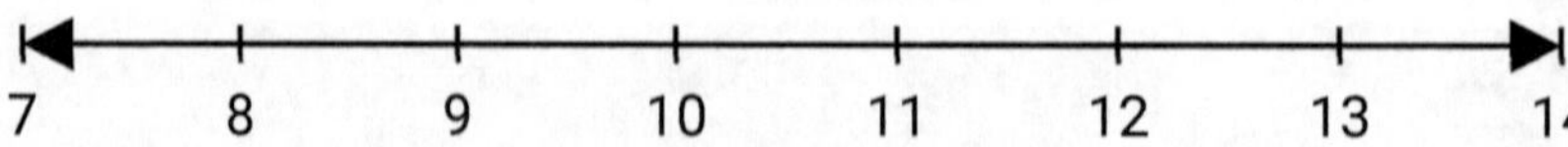

3.

$$x + 5 < -3$$

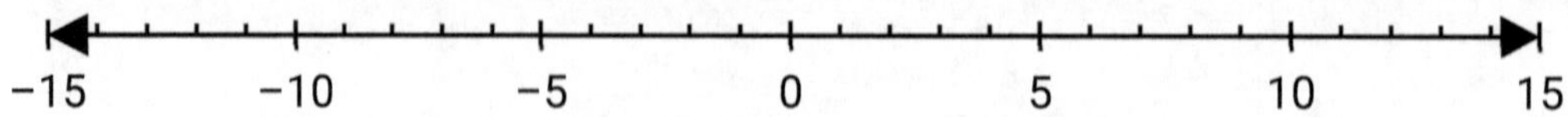

4.

$$-2x \leq 16$$

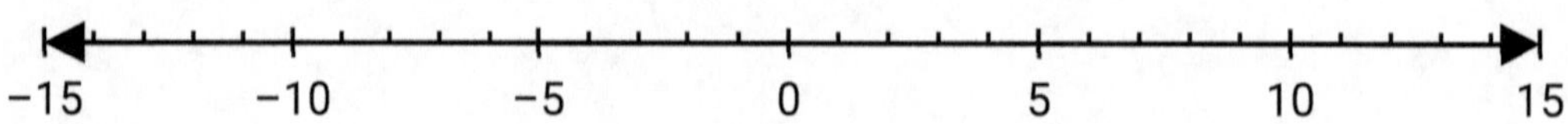

5.

$$\frac{x}{4} \leq -2$$

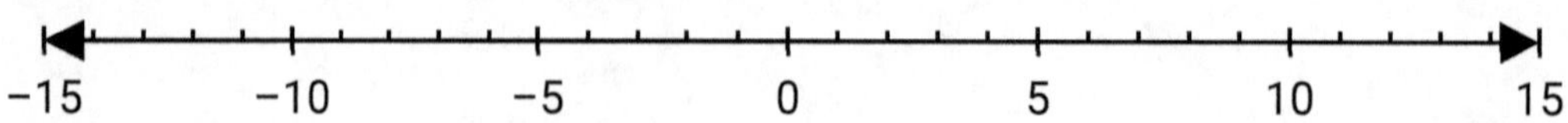

Day 18

Determine the Constant of Proportionality

Determine the constant of proportionality from each ordered pair.

1. $(2, 10)$ 2. $(4, 28)$ 3. $(6, 15)$ 4. $(9, 27)$

__________ __________ __________ __________

Determine the constant of proportionality from each table.

5.

x	y
1	4
3	12
5	20

6.

x	y
2	3
6	9
10	15

7.

x	y
2	1
4	2
6	3

8.

x	Y
−1	5
2	−10
−5	20

__________ __________ __________ __________

Determine the constant of proportionality from each equation.

9. $y = 7x$ 10. $y = \frac{1}{2}x$ 11. $y = -3x$ 12. $4y = 20x$

__________ __________ __________ __________

Day 19

Proportional or Non-Proportional Relationships

Determine if each relationship is proportional or non-proportional.

1. $y = 3x$

____

2. $y = 2x + 5$

____

3.

x	y
1	4
2	8
3	12

____

4.

x	y
2	5
4	10
6	15

____

5. $y = 7x - 1$

____

6. $y = \frac{1}{2}x$

____

Day 20

Simplifying Two-Variable Equations

Simplify the following equations to the form $y = mx + b$.

1. $$3x + 4y + 2x = 20$$

2. $$6x + 2y - 3x = 30$$

3. $$5x - 2y + 3x + y = 18$$

4. $$8x - 3y + 4x - y = 24$$

5. $$10x + 5 - 2x + 3y = 29$$

6. $$4x + 7y - 2x + 3y + 2 = 22$$

7. $$9x - 4 + 2x + 6y = 50$$

8. $$2x + 3y + 5x - y = 14 - 2x$$

9. $$3(x + 2y) = 21 - 3$$

10. $$4(2x - y) + 3x = 20$$

Day 21

Graphing Two-Variable Equations

Finish the table and make a graph for each equation.

1. $y = 2x$

x	y
−2	
−1	
0	
1	

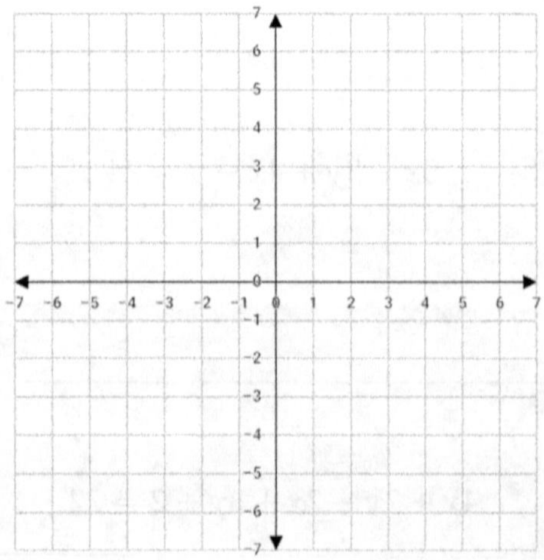

2. $y = x + 3$

x	y
−2	
−1	
0	
1	

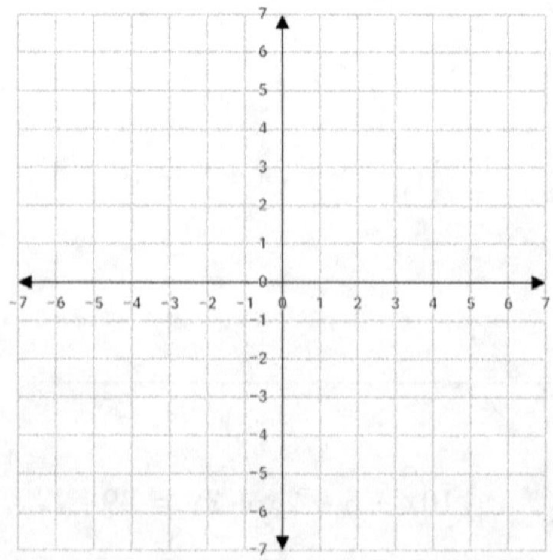

3. $y = -3x - 2$

x	y
−2	
−1	
0	
1	

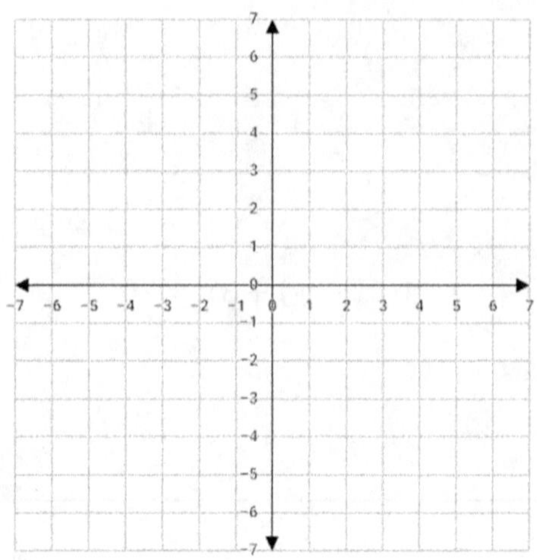

4. $y = \frac{1}{2}x$

x	y
−2	
−1	
0	
1	

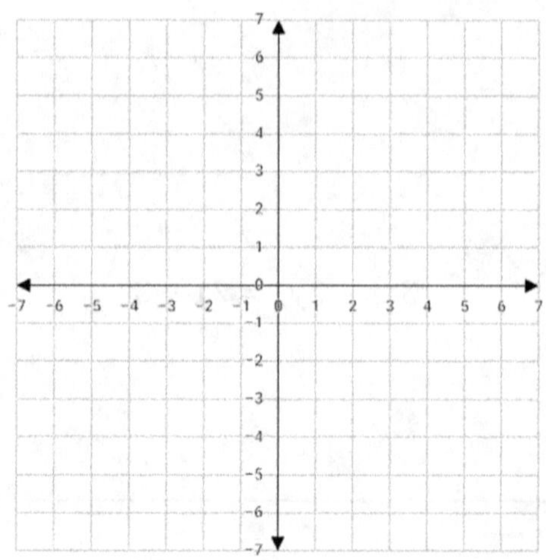

Day 22

Solving using Substitution

Solve the following equations using substitution.

1. $x = 3$

$2x + 3y = 24$

____

2. $y = 2$

$5x - y = 18$

____

3. $x = 2$

$3x + 2y - 4 = 12$

____

4. $y = 7$

$4x - 3y = -17$

____

5. x= 5

$x^2 + y = 28$

____

6. $y = 2$

$2(x + y) = 16$

____

7. $x = -2$

$3x - 2y + 4 = -12$

____

8. $y = -7$

$4x - y = 19$

____

Chapter 3: Proportional Reasoning

A **ratio** expresses the relationship between two values. For example, a veterinarian sees 35 dogs and 25 cats per week. The ratio of dogs to cats as 35 to 25, 35 : 25, or $\frac{35}{25}$. You can simplify a ratio by finding the greatest common factor and dividing each term which creates an equivalent ratio, such as $\frac{7}{5}$.

A **rate** is a comparison of two values with different units. Some common examples are the comparison of dollars to hour, miles to hours, and beats to seconds. Rates can be expressed in a variety of ways, including using language, as quotients, using equivalent rates, algebraically, using tables, or graphically. A **unit rate** is a rate written in simplest form, where one of the two units is one.

We can represent ratios and rates using proportions and scale factors. A **proportion** is an equation with two equivalent ratios. We use proportions to compare ratios and to solve ratio and rate problems. Suppose a recipe requires 2 cups of sugar and 3 cups of flour. This can be described by the ratio 2 : 3, or $\frac{2}{3}$. Now, what if we were making a recipe with 8 cups of sugar? We can use proportions to help us solve this problem where x is the number of cups of flour needed.

$$\frac{2}{3} = \frac{8}{x}$$

Proportions are written in the form:

$$\frac{a}{b} = \frac{c}{d}$$

They can also be written in the form $ad = cb$ when the equation is rearranged.

$$\begin{aligned} d \times \frac{a}{b} &= \frac{c}{d} \times d \\ b\left(d \times \frac{a}{b}\right) &= c \times b \\ d \times a &= c \times b \\ ad &= cb \end{aligned}$$

This is also known as **cross multiplying**. When we have a proportion, we can cross multiply to solve.

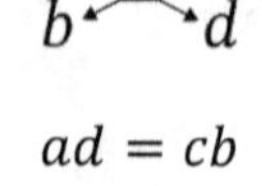

$$ad = cb$$

Now, solve the proportion for the number of cups of flour needed for 8 cups of sugar. So, if we are using 8 cups of sugar, we need 12 cups of flour.

$$\frac{2}{3}=\frac{8}{x}$$
$$2\times x=8\times 3$$
$$\frac{2x}{2}=\frac{24}{2}$$
$$x=12$$

These are equivalent ratios.

$$\frac{2}{3}=\frac{8}{12}$$

There are multiple different kinds of percent proportion problems. An easy way to solve percent proportions is by remembering a simple set up:

$$\frac{Part}{Whole}=\frac{\%}{100}$$

Then, you solve for the missing piece of the proportion usually starting with cross multiplication.

Similar figures are figures that have the same shape, but are different sizes. Similar figures have the same number of sides and angles. Their corresponding angles are congruent, and their sides are proportional. Similar figures are the same shapes but different sizes. The change between similar figures can be described as a scale factor. The **scale factor** is the ratio between corresponding side lengths. It determines how big or small the new shape is from the original shape.

A **scale drawing** is a drawing used to represent similar figures. A map is an example of a scale drawing. Maps represent distances and measurements that are much too large to fit on a wall or sheet of paper, and the distance between locations on a map is proportional to the distance between those locations on earth. A scale drawing is a proportional representation, which means that the real measurements are proportional to the drawn measurements. A scale is used to define this proportional relationship.

The **scale** in a scale drawing is included in the key and shows the relationship between the original item and the scale drawing. The scale includes a measurement from the scale drawing and gives an equivalent measurement from the original item. An example of a scale is 1 inch = 100 miles. This means that in the scale drawing, 1 inch is equivalent to 100 miles in the original.

Lists and tree diagrams can be used to represent sample spaces for simple events. A **simple event** is an event where there is one trial. For example, a coin flip is a simple event, since it has one trial that has one outcome. A **sample space** is a representation of all possibilities in an event.

The **probability** of an outcome describes the comparison between a proposed outcome and the total possible outcomes of an event. When performing calculations involving probability, these percentages are usually expressed as decimals or fractions. For instance, a six-sided dice is rolled. The probability that the dice will land on the number 1 is $\frac{1}{6}$ because there is one side that reads 1 and there are six total sides.

Complete one worksheet each day in this chapter. Each day should take no longer than 15 minutes to complete.

Day 23

Ratios and Rates

Complete each problem below.

1. There are 3 red marbles, 8 green marbles, and 12 blue marbles in a bag. Write the ratio of red to blue marbles in simplest form.

____

2. A class has 15 boys and 10 girls. What is the ratio of boys to girls?

____

3. Simplify the ratio 24 : 42.

____

4. A car travels 180 miles in 3 hours. What is the unit rate in miles per hour?

____

5. A store sells 5 notebooks for $10. What is the cost per notebook?

____

6. 72 students are in 6 classes. What is the unit rate of students per class?

____

7. A runner completes 10 miles in 2 hours. What is the unit rate?

____

8. Liam got 9 apples for $4.50. How much did he spend per apple?

____

9. A faucet fills 15 gallons in 5 minutes. What is the rate per minute?

____

10. A printer prints 30 pages per minute. How many pages will be printed after 4 minutes?

____

Day 24

Proportions

Solve the following proportions.

1. $\frac{x}{6} = \frac{4}{12}$

2. $\frac{5}{8} = \frac{x}{24}$

3. $\frac{9}{3} = \frac{x}{7}$

4. $\frac{12}{4} = \frac{6}{x}$

5. $\frac{3}{5} = \frac{18}{x}$

6. $\frac{7}{9} = \frac{x}{27}$

7. $\frac{15}{x} = \frac{3}{4}$

8. $\frac{2}{3} = \frac{x}{15}$

9. $\frac{4}{10} = \frac{x}{25}$

10. A store has a sale to get 3 pencils for $1.50. How much would it cost to purchase 9 pencils?

DAY 25

PERCENT PROPORTIONS – FINDING THE PART

Write a proportion for each problem and then solve.

1. What is 25% of 80?

____

2. What is 10% of 300?

____

3. What is 15% of 60?

____

4. What is 40% of 90?

____

5. What is 12% of 200?

____

6. What is 5% of 120?

____

7. Grace bought a dress at a store having a 30% off sale. The original tag on the dress was $46. How much money did Grace save on the dress?

____

8. Nick got a 4% raise at his job. He made $86,000 annually before the raise. How much more money will he receive every year?

____

9. Katie went to eat sushi. Since it was her birthday, they gave her a 25% discount on her food. The price before the discount was $24. How much did she end up paying after the discount was added?

____

10. Samuel went out to eat with his friends. His bill was $18, and he wanted to leave a 20% tip for the waitress. How much will his bill be after he includes tip?

____

DAY 26

PERCENT PROPORTIONS – FINDING THE WHOLE

Write a proportion for each problem and then solve.

1. 20 is 25% of what number?

____

2. 15 is 30% of what number?

____

3. 36 is 60% of what number?

____

4. 18 is 15% of what number?

____

5. 27 is 90% of what number?

____

6. 84 is 70% of what number?

____

7. Eric paid $46,000.30 for a new truck. He got 15% off the original price. What was the original price of the truck?

____

8. Peter made 80% of his shots in a basketball game. If he made 24 shots, how many did he attempt total?

____

9. Jake gave each kid 3% of his bonus as a present. If each kid received $80, how much was Jake's entire bonus?

____

10. Jacob went to work out on a trail. He ran 75% of the time and walked the rest. If he ran for 4.5 miles, how many total miles did he go?

____

Day 27

Percent Proportions – Finding the Percent

Write a proportion for each problem and then solve.

1. 16 is what percent of 80?

____

2. 18 is what percent of 60?

____

3. 12 is what percent of 48?

____

4. 21 is what percent of 70?

____

5. 45 is what percent of 60?

____

6. 18 is what percent of 90?

____

7. Karson gave each of his 3 kids a slice of pie. There were 10 total slices. What percentage of the pie is left?

____

8. Harley has 5 horses. She has cleaned 12 hooves so far. What percentage of the hooves has she cleaned so far?

____

9. A company ordered food for 29 out of 36 employees. What percentage of employees did not get any food?

____

10. Jamie is on a payment plan to pay for her new couch. So far, she has made 5 out of 12 payments. What percentage has she paid of the plan?

____

Day 28

Similar Figures

Use proportions to help find the missing values on the similar figures below.

1. ΔABC is similar to ΔDEF. Find the value of x.

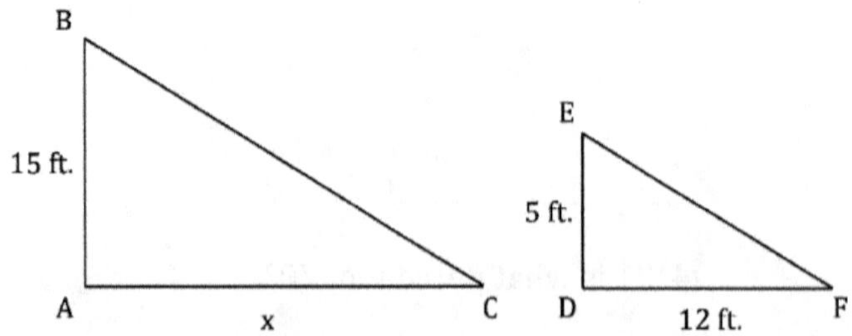

2. Trapezoid $QMNP$ is similar to trapezoid $URST$. What is the value of x?

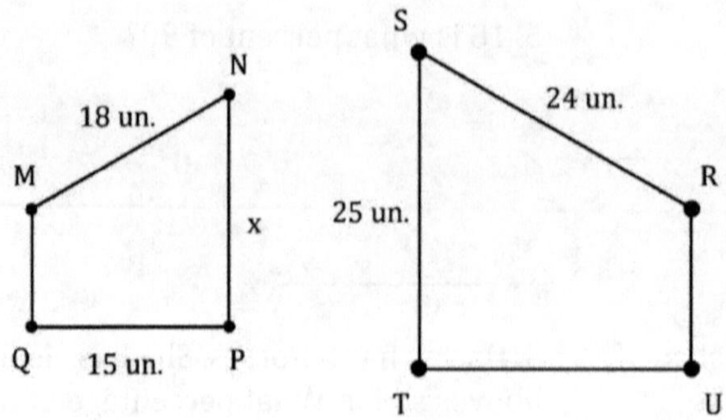

3. Triangle ABC is similar to trianglePQR. Find the length of $\overline{AC}$.

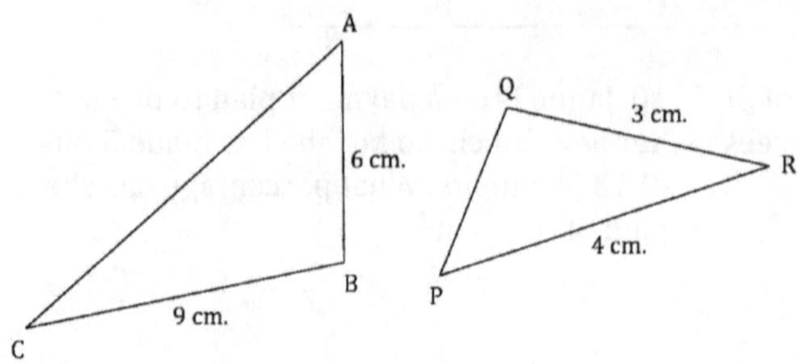

DAY 29

SCALE DRAWINGS

Use proportions to solve with the given scale factors.

1. The scale factor is $\frac{1\text{ inch}}{10\text{ feet}}$. How many feet long is the bus?

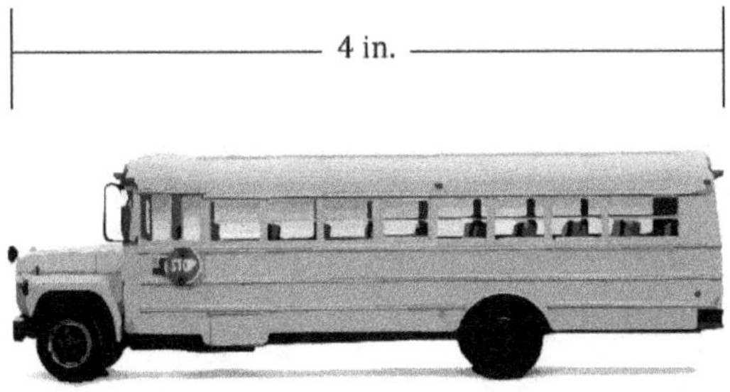

____

2. The scale factor is $\frac{1\text{ cm}}{7\text{ in}}$. Determine the area of the TV in square inches.

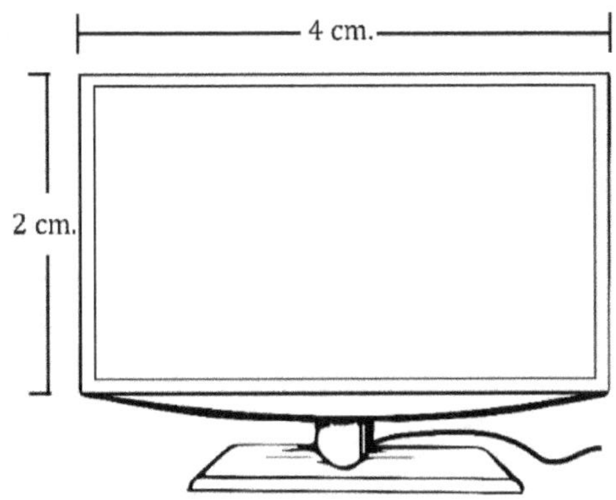

____

3. The scale factor is $\frac{1\text{ in}}{4\text{ ft}}$. How many feet long is the kitchen?

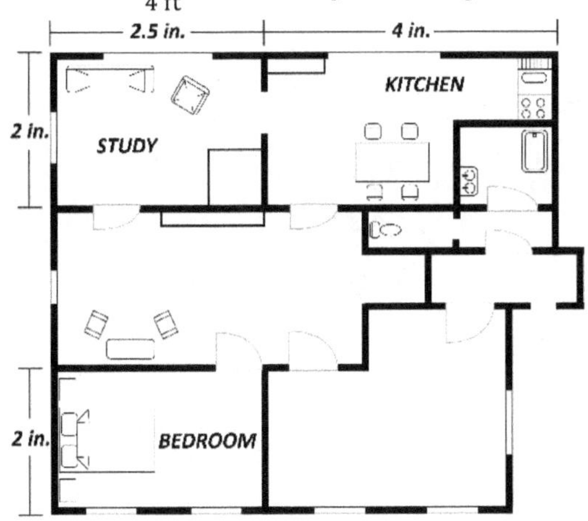

____

Day 30

Dilation on a Coordinate Plane

Determine the scale factor of the dilation from figure A to figure B in each question below.

1.

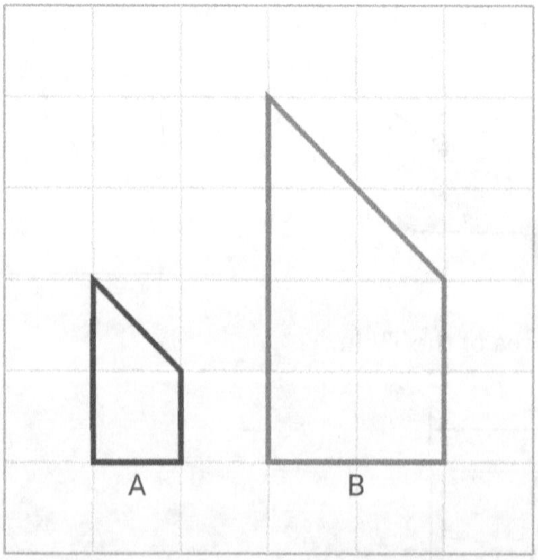

____

2.

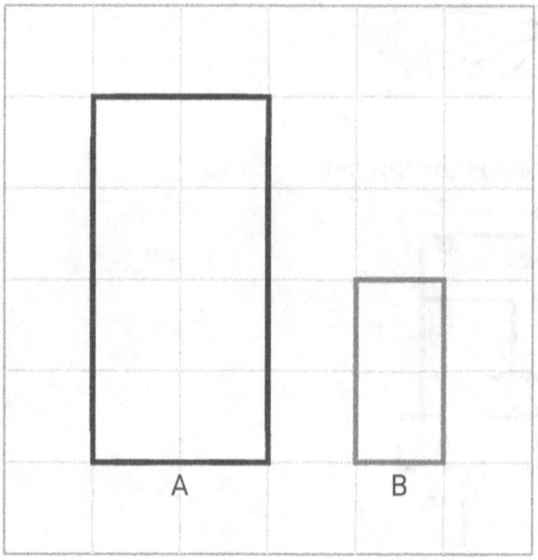

____

Day 31

Sample Spaces – Lists, Tables, and Tree Diagrams

Answer the following questions.

1. Henry rolls two six-sided dice. Use an organized list to represent the sample space of dice rolls. Determine the total number of possible dice rolls.

____

2. A class had a cookout at the end of the school year. Each class member could choose one of the following entrées: hot dog, hamburger, or barbecue sandwich. They could also choose one of the following sides: chips, potato salad, or side salad. Suppose a student chose their entrée and side at random. Use a sample space table to represent the sample space of the possible combinations of orders. Determine the total number of unique orders a student can place at the cookout.

____

3. Enrique flips a fair coin three times to see how many times the coin will land on heads. Draw a tree diagram to represent the sample space. Determine how many different combinations of coin flips can occur.

____

Day 32

Simple Probability

Find the probability of each problem below.

1. Rolling a 3 on a die.

____

2. Flipping heads on a coin.

____

3. Rolling an even number on a die.

____

4. Picking red from a bag with 3 red and 2 blue.

____

5. Picking blue from a bag with 4 blue and 6 green.

____

6. Rolling an odd number on a die.

____

7. Flipping tails on a coin.

____

8. Rolling a number greater than 4 on a die.

____

9. Picking green from a bag with 2 green and 8 yellow.

____

10. Picking black from a bag with 5 black and 5 white.

____

Chapter 4: Geometry

The **area** of a shape describes the total measure of the amount of space inside the polygon. We describe area using **square units** because area can be described as the number of squares that cover the inside of a shape. We can also use formulas to find the area of a different shapes.

Parallelogram	A parallelogram is a quadrilateral with opposite sides that are both congruent and parallel.	$\boldsymbol{Area = bh}$ 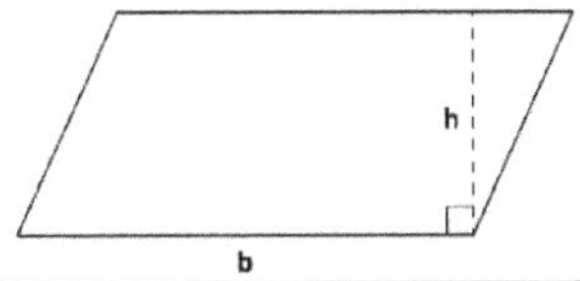
Rectangle	A rectangle is a special kind of parallelogram that has all right angles.	$\boldsymbol{Area = bh}$ 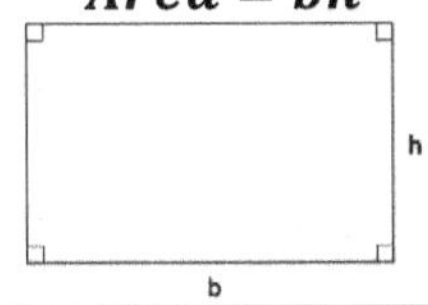
Trapezoid	A trapezoid is a quadrilateral with at least one pair of parallel sides. The parallel sides are called the bases of the trapezoid. To differentiate the bases, the following notation is used: $b_1 = base\ 1, b_2 = base\ 2$.	$\boldsymbol{Area = \frac{(b_1 + b_2)}{2} \cdot h}$ 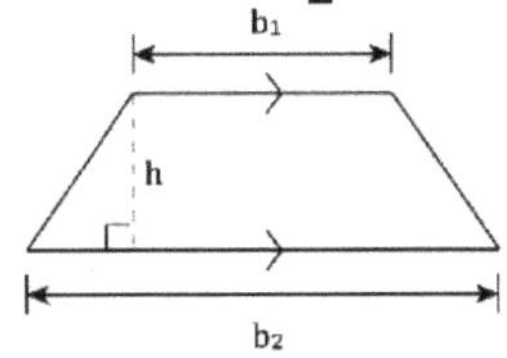
Triangle	A triangle is a three-sided polygon with three vertices.	$\boldsymbol{Area = \frac{1}{2}(bh)}$ 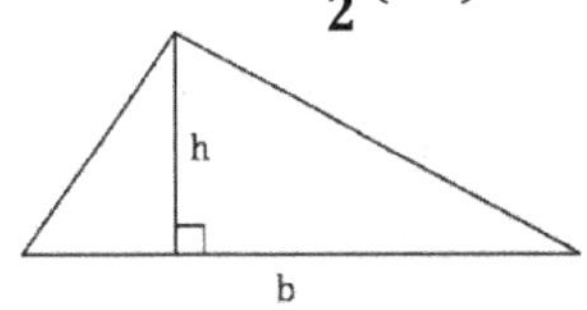

A **composite shape,** or **composite figure,** is a shape that is made up of two or more two-dimensional shapes. The figures below are all composite figures.

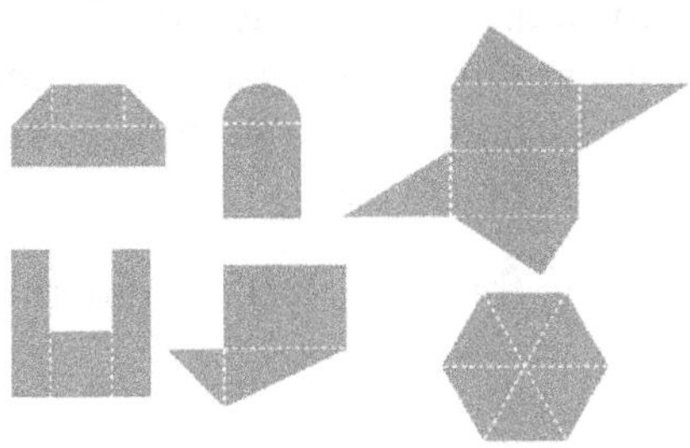

Composite figures can be split or decomposed into component shapes to calculate area.

The relationships between the measurements of a circle are unique. A circle does not have any sides or angles. The perimeter of a circle is called the **circumference**, and the width of a circle is called the **diameter**. The **radius** of a circle is the length from the center to any point along the circumference. The radius is half of the diameter.

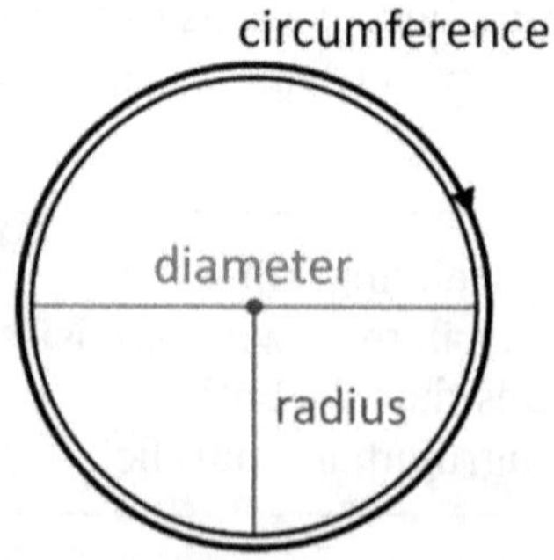

Pi, represented by the symbol π, is the constant for the proportional relationship between a circle's circumference and diameter. Pi is often rounded to 3.14. Formulas are used to find the circumference and area of circles and they all involve π.

The formula to solve for the circumference of a circle is $C = \pi d$ where C represents the circumference and d represents the diameter. The formula for the area of a circle is $A = \pi r^2$ where A represents the area and r represents the radius.

The **surface area** of a three-dimensional figure is the total two-dimensional area of the surface of the figure. We can find the surface area of a three-dimensional figure by adding the area of each face of the figure.

A **net** is an "unfolded" version of a three-dimensional figure. It is as if the figure is opened and laid flat. The net of a three-dimensional figure accurately identifies the polygon faces that make a three-dimensional figure.

Volume is the measurement of space inside a three-dimensional shape or object. We use **cubic units**, which can be thought of as cubed units, to describe volume because volume is the number of unit cubes it takes to fill a space.

A **prism** is a three-dimensional shape with two equivalent bases, or faces, that are parallel. A rectangular prism has two equivalent rectangular (including square) bases that are parallel. A triangular prism has two equivalent triangular bases that are parallel. To find the volume of a prism, V, we multiply the area of one base, B, by the height of the prism, h. This relationship is reflected by the formula $V = Bh$. The "base" of a two-dimensional shape is represented by a lower case b, where the "base" of a three-dimensional shape is represented by an upper case B. The base of a three-dimensional shape is the two-dimensional area of the shape that forms the base.

Complete one worksheet each day in this chapter. Each day should take no longer than 15 minutes to complete.

Day 33

Identifying Parts of a Circle

Identify the parts of the circle.

1.

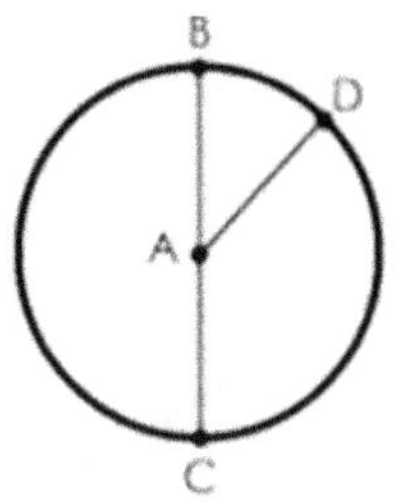

Circle: ___A___

Radius: ___AB, AC, A___

Diameter: ___BC___

2.

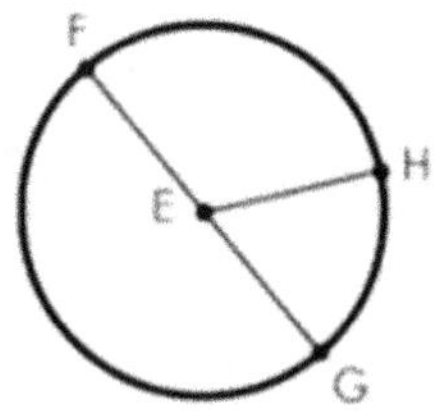

Circle: ____________

Radius: ____________

Diameter: __________

3.

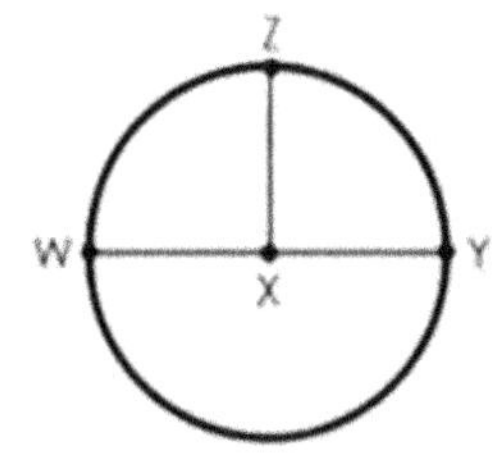

Circle: ____________

Radius: ____________

Diameter: __________

4.

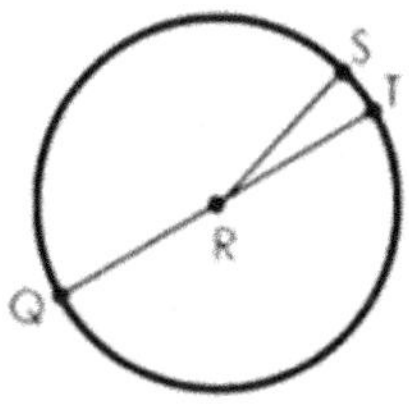

Circle: ____________

Radius: ____________

Diameter: __________

5.

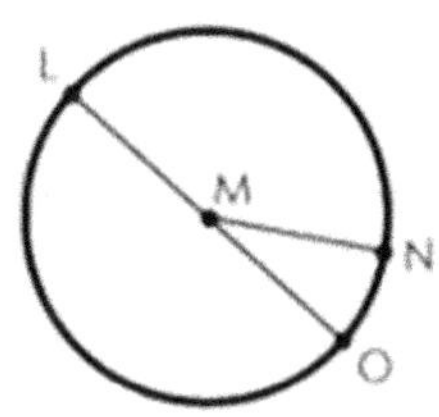

Circle: ____________

Radius: ____________

Diameter: __________

6.

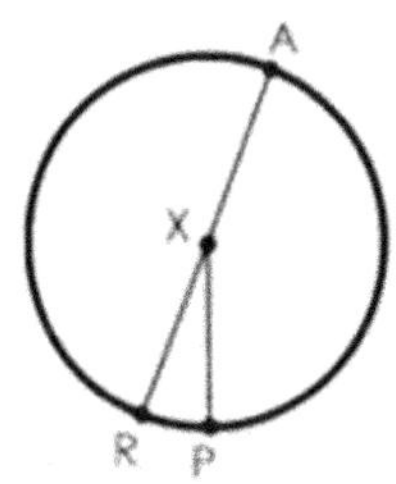

Circle: ____________

Radius: ____________

Diameter: __________

Day 34

Finding the Radius of a Circle

Use the diameter to determine the length of the radius of each circle. Write out the entire equation. Don't forget to include the units in your answer.

1.

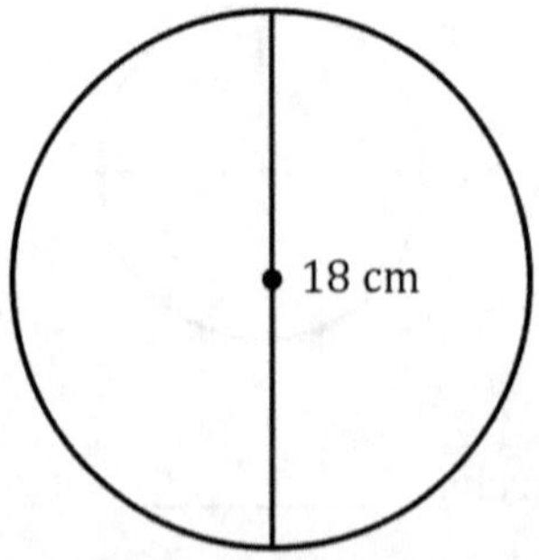

Radius: ____________________

2.

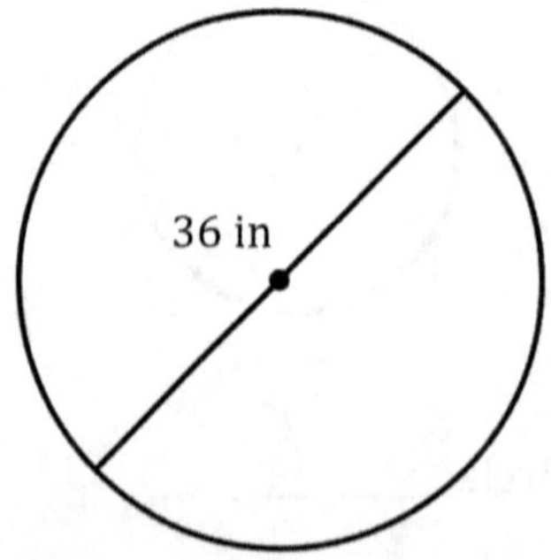

Radius: ____________________

3.

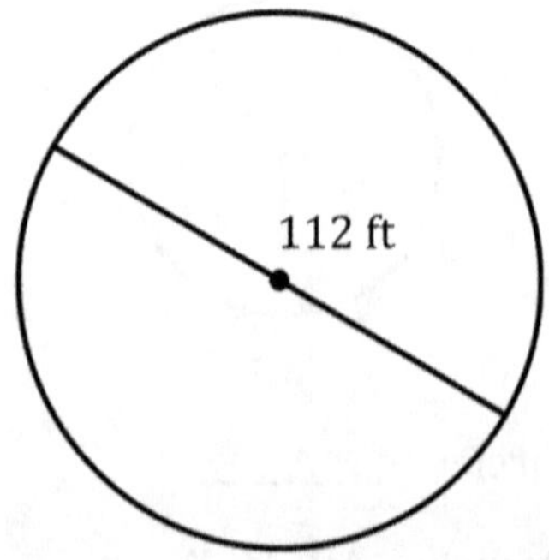

Radius: ____________________

4.

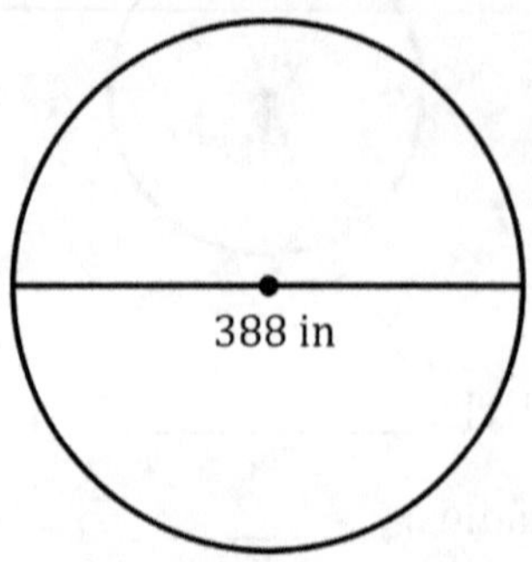

Radius: ____________________

5.

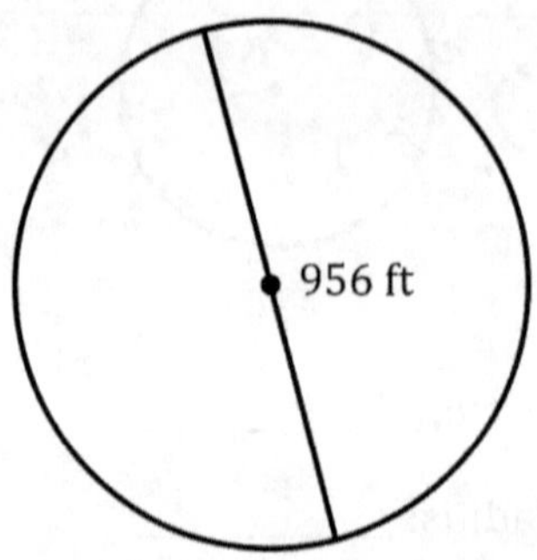

Radius: ____________________

6.

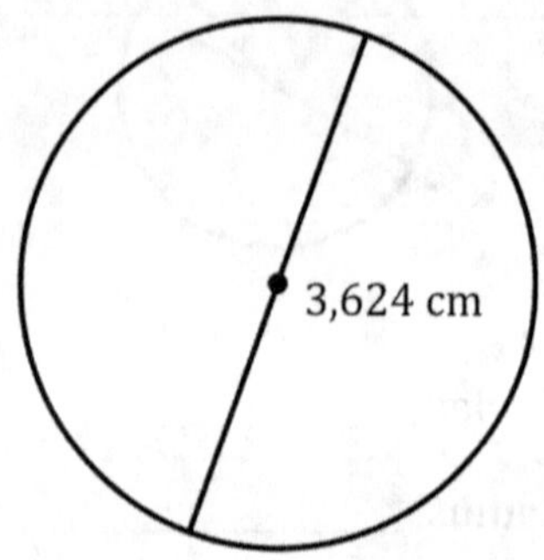

Radius: ____________________

Day 35

Finding the Diameter of a Circle

Use the radius to determine the length of the diameter of each circle. Write out the entire equation. Don't forget to include the units in your answer.

1.

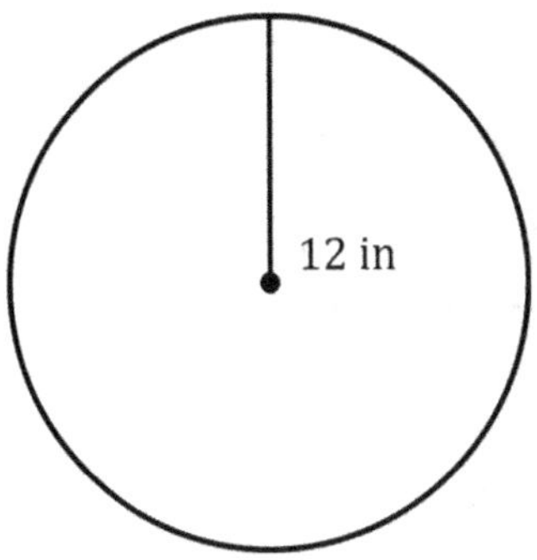

Diameter: ______________

2.

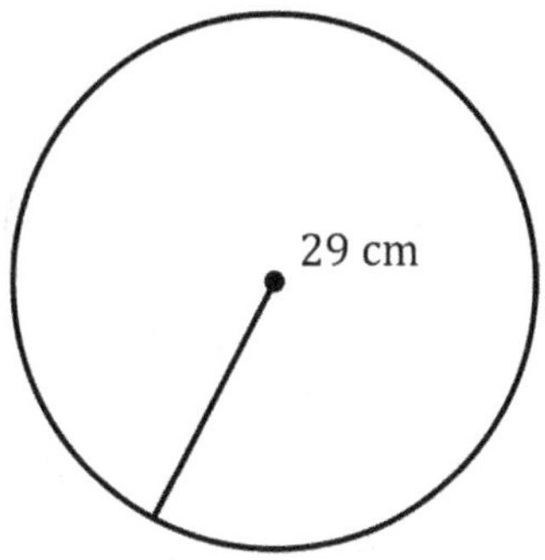

Diameter: ______________

3.

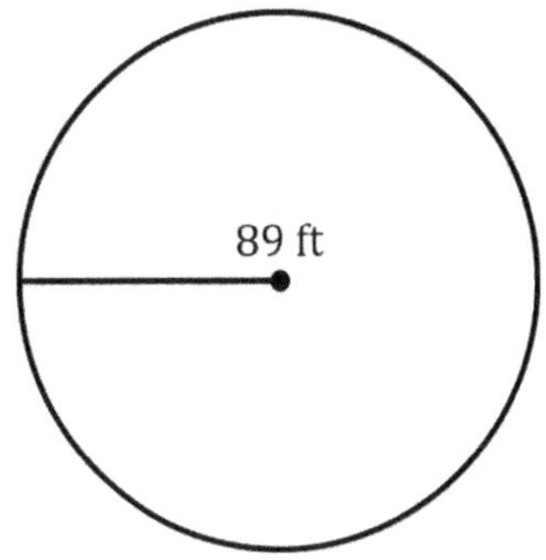

Diameter: ______________

4.

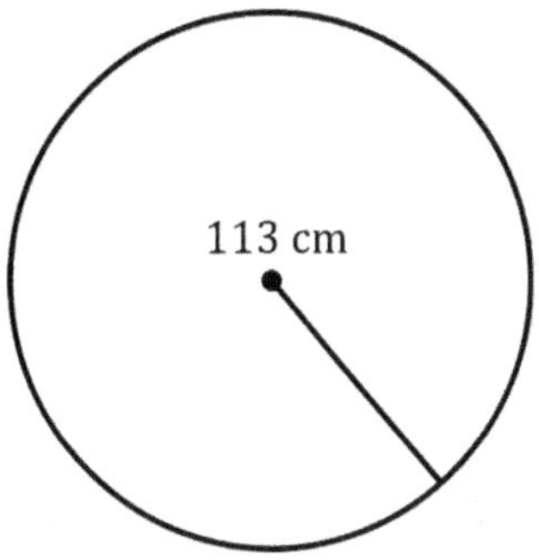

Diameter: ______________

5.

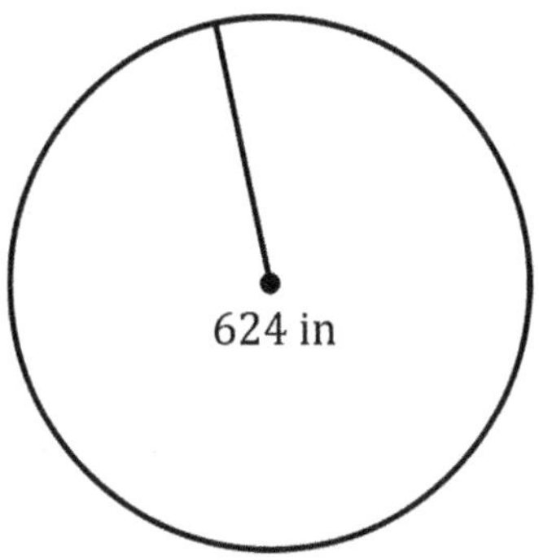

Diameter: ______________

6.

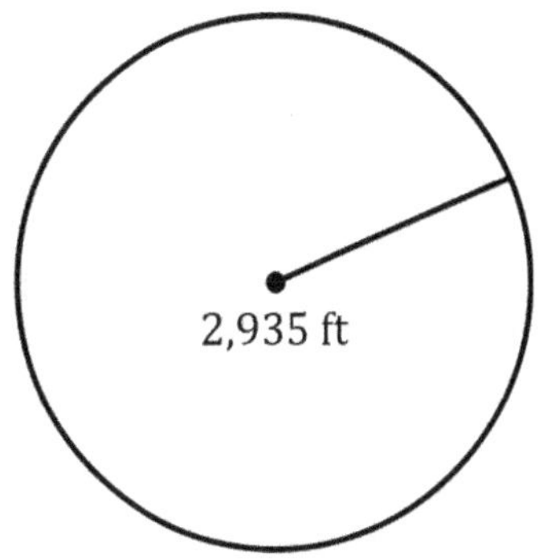

Diameter: ______________

DAY 36

FINDING THE CIRCUMFERENCE OF A CIRCLE

Determine the circumference of the circle. Write out the entire equation. Don't forget to include the units in your answer.

1.

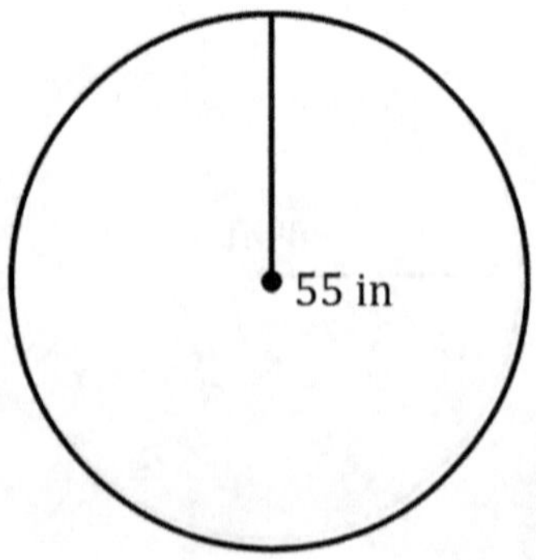

Circumference:

____

2.

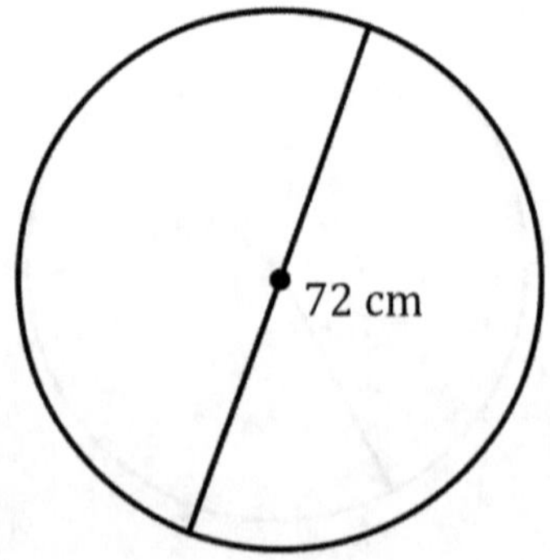

Circumference:

____

3.

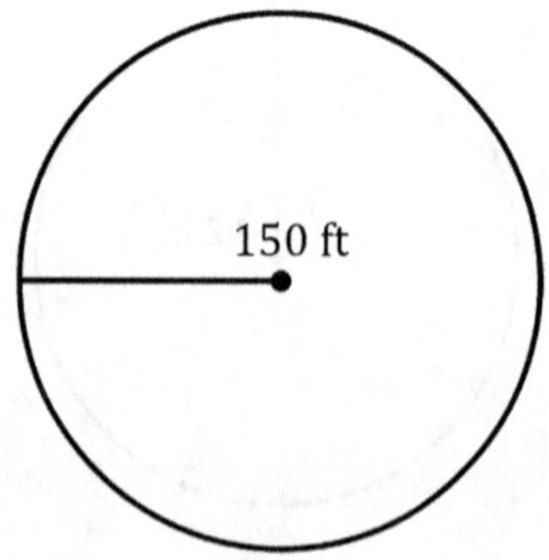

Circumference:

____

4.

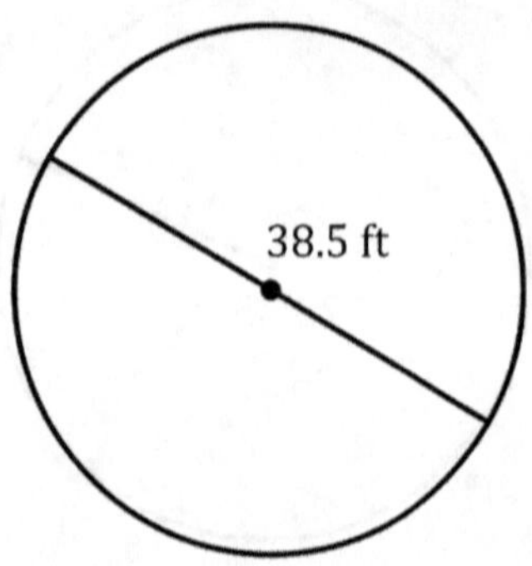

Circumference:

____

5.

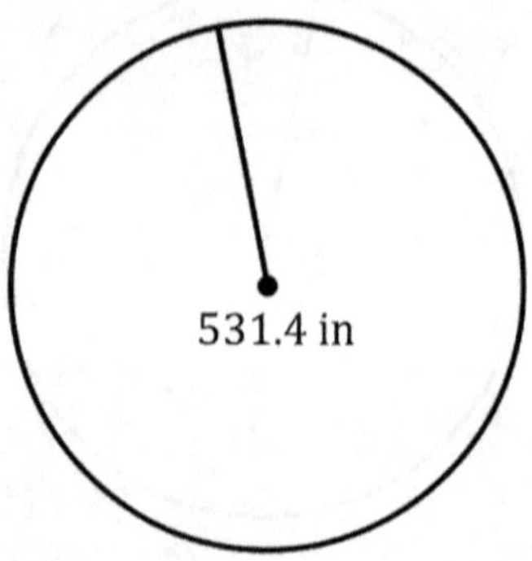

Circumference:

____

6.

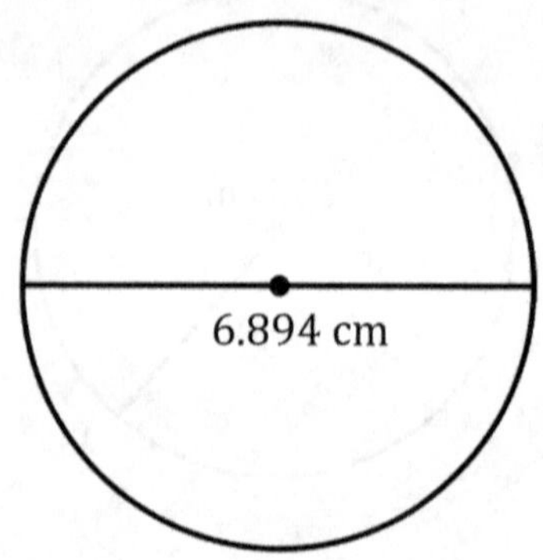

Circumference:

____

DAY 37

AREA OF A CIRCLE

Determine the area of the circle. Write out the entire equation. Don't forget to include the units in your answer. Round your answer to the nearest hundredth.

1.

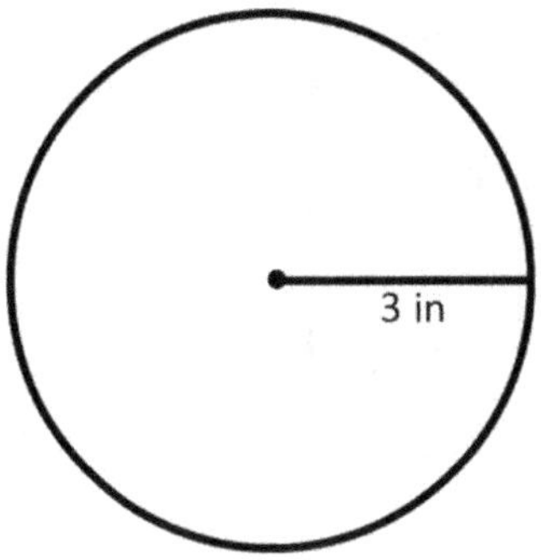

Area:

____

2.

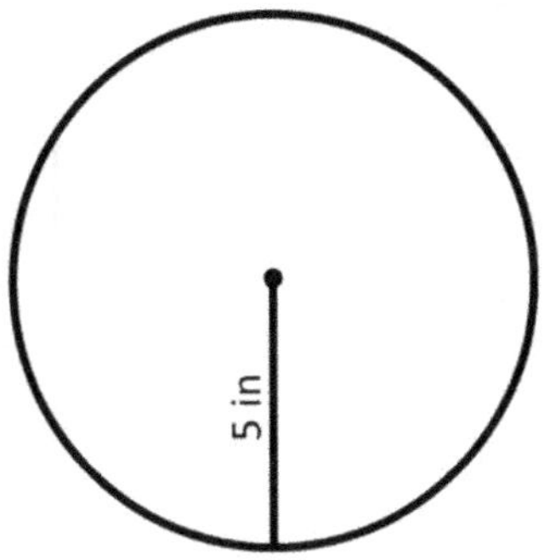

Area:

____

3.

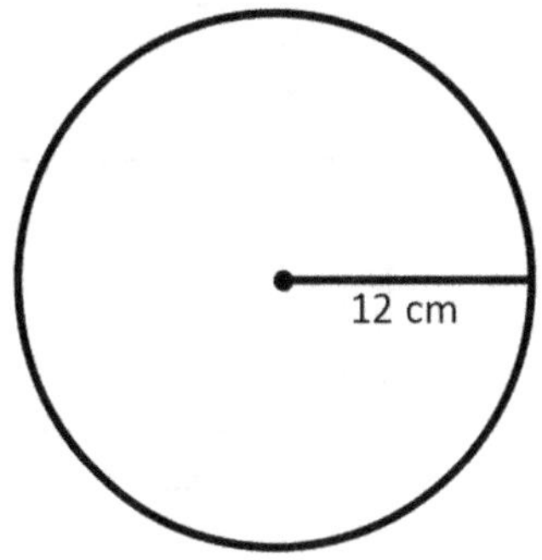

Area:

____

4.

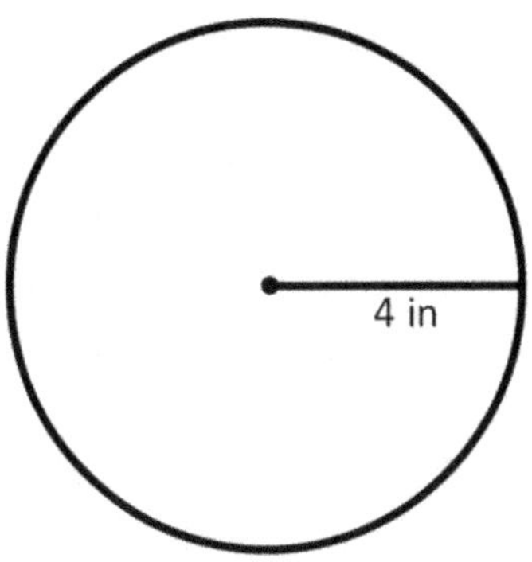

Area:

____

5.

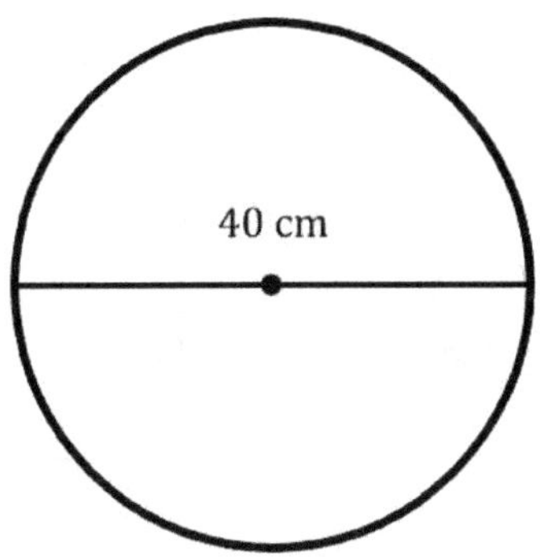

Area:

____

6.

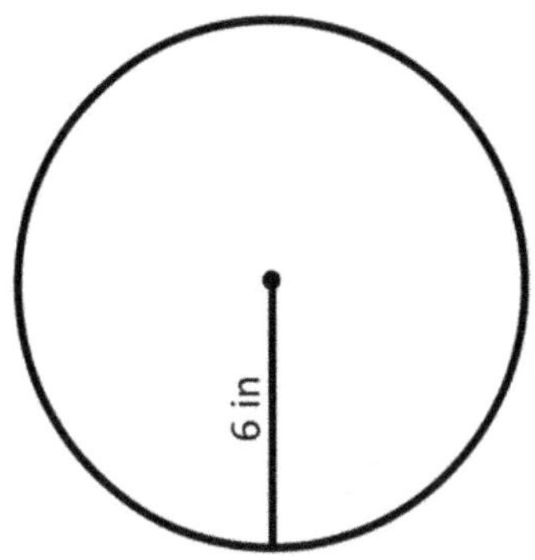

Area:

____

Day 38

Surface Area and Volume of Rectangular Prisms

Determine the surface area and volume of each shape. Don't forget to include the units in your answer.

1.

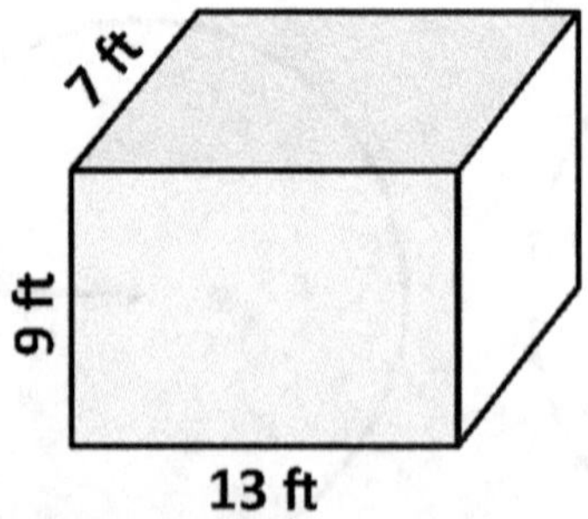

Surface Area: ____________________

Volume: ________________________

2.

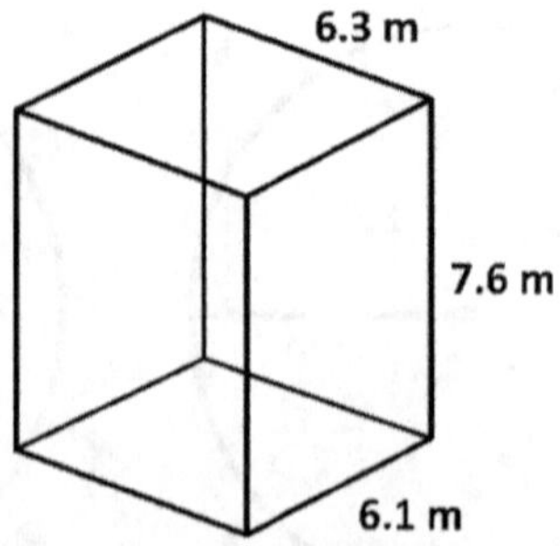

Surface Area: ____________________

Volume: ________________________

3.

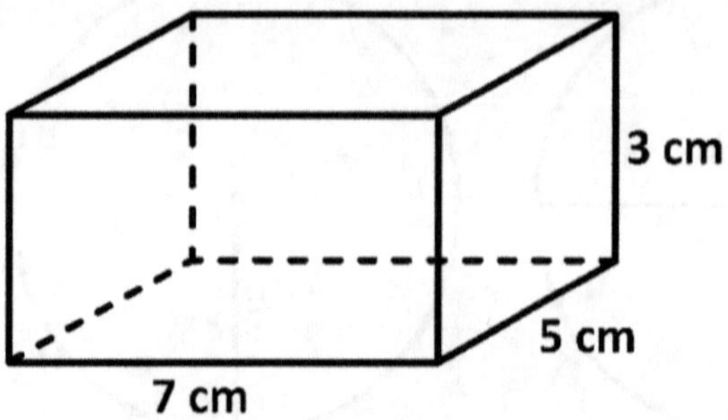

Surface Area: ____________________

Volume: ________________________

4.

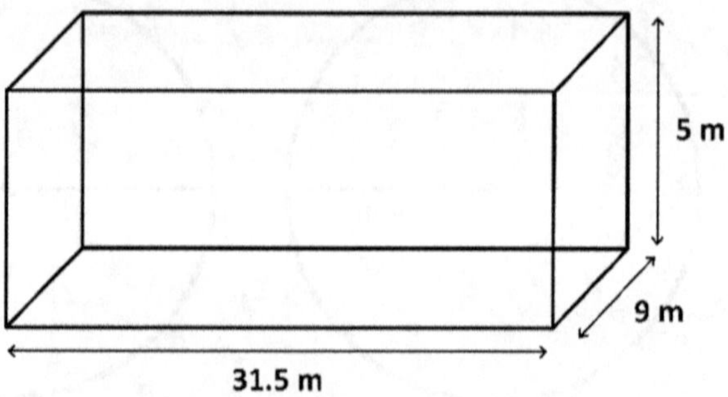

Surface Area: ____________________

Volume: ________________________

DAY 39

VOLUME OF TRIANGULAR PRISMS

Determine the volume of each shape. Don't forget to include the units in your answer.

1.

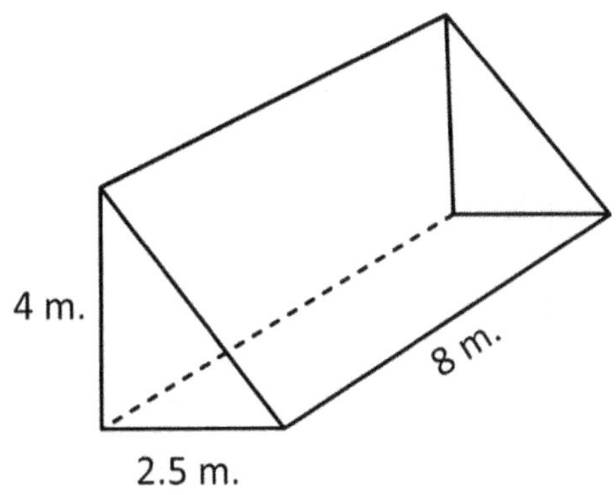

Volume: ______________________

2.

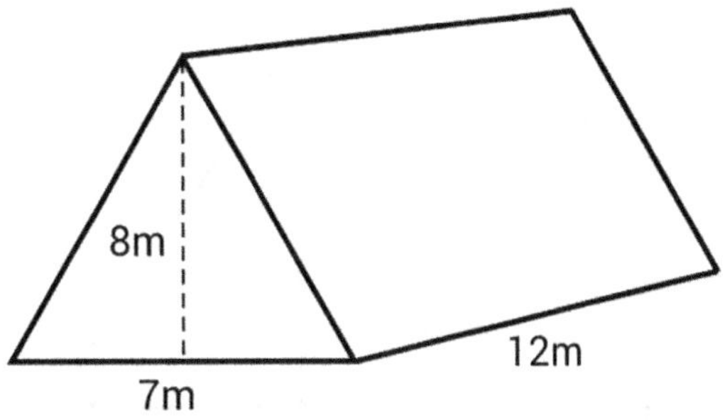

Volume: ______________________

3.

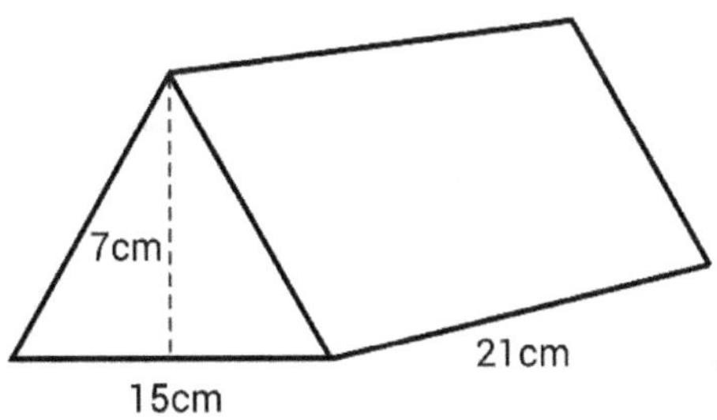

Volume: ______________________

4.

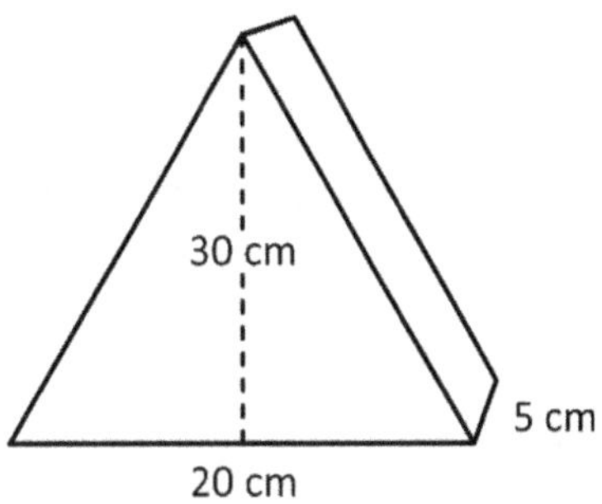

Volume: ______________________

Day 40

Surface Area of Triangular Prisms

Determine the surface area of each shape. Don't forget to include the units in your answer.

1.

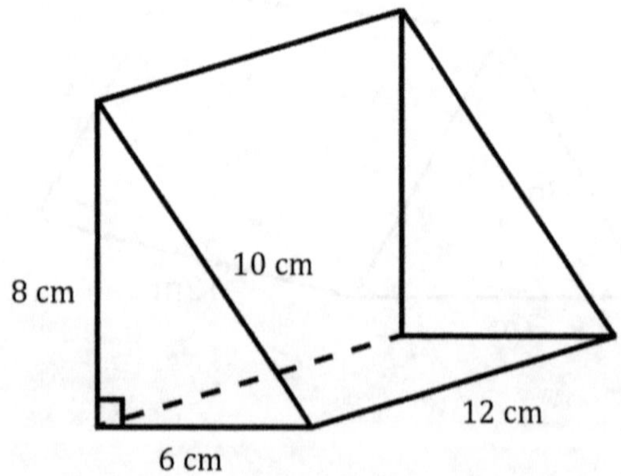

Surface Area: ____________________

2.

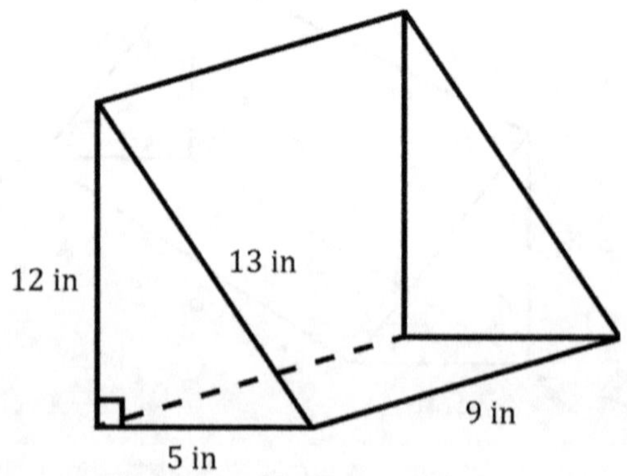

Surface Area: ____________________

3.

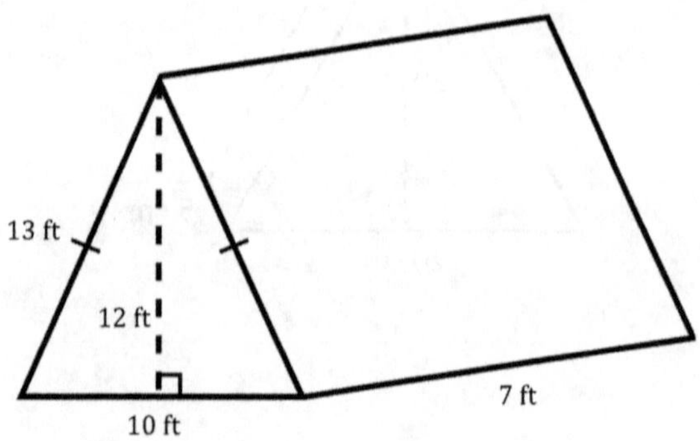

Surface Area: ____________________

4.

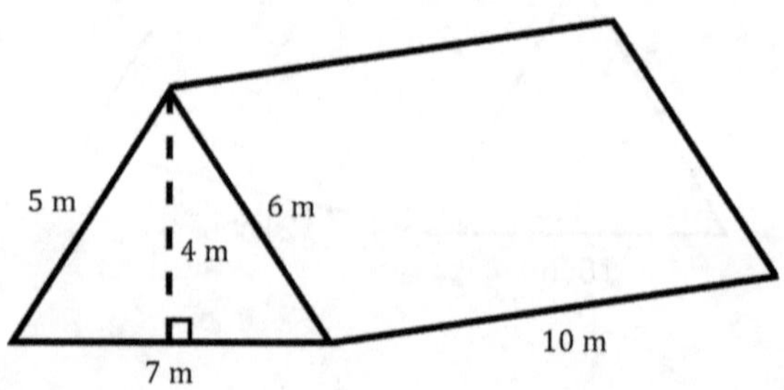

Surface Area: ____________________

Day 41

Surface Area of Pyramids

Determine the surface area of each shape. Don't forget to include the units in your answer.

1.

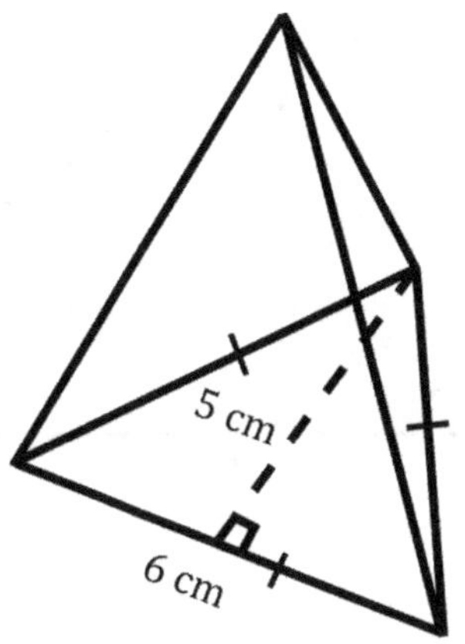

Surface Area: ____________________

2.

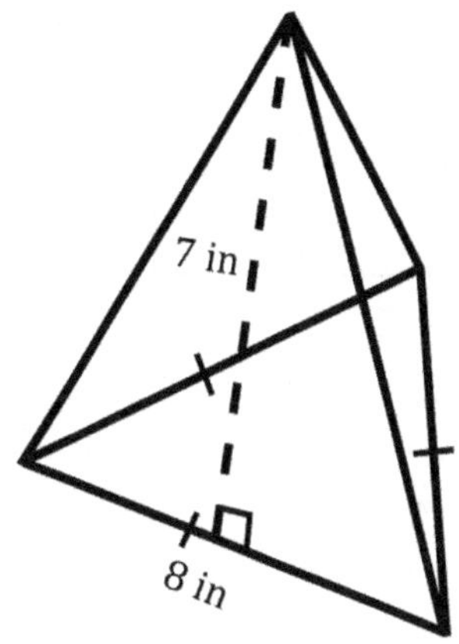

Surface Area: ____________________

3.

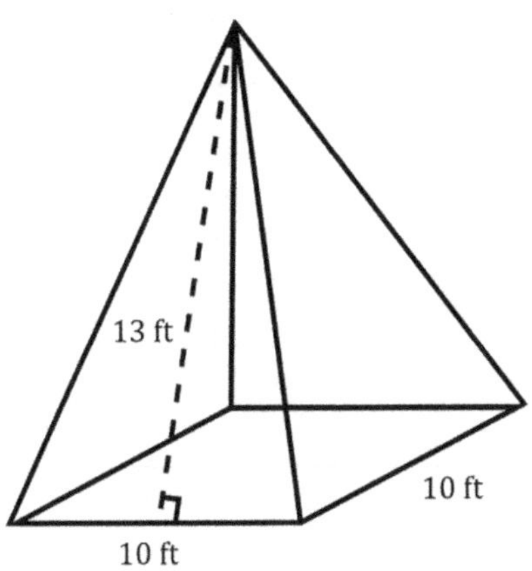

Surface Area: ____________________

4.

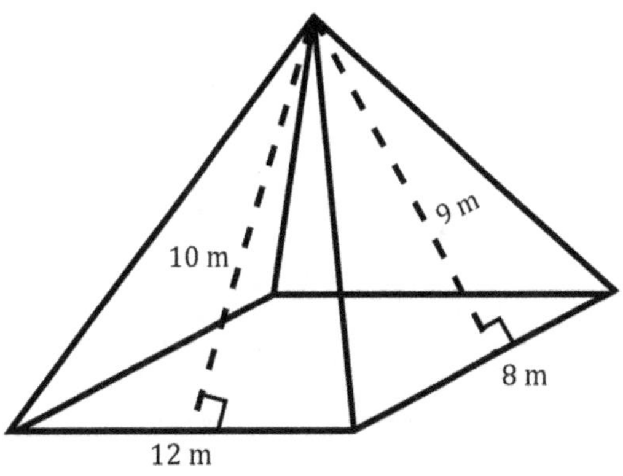

Surface Area: ____________________

Day 42

Area of Composite Figures

Determine the area of the figures below. Don't forget to include the units in your answer. Round your answer to the nearest hundredth.

1.

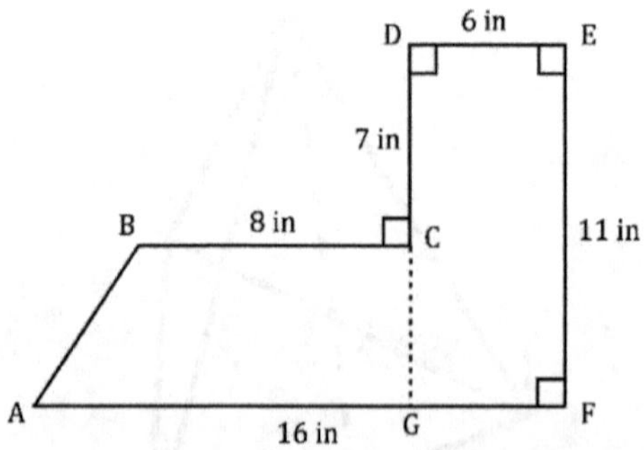

Area: ______________________

2.

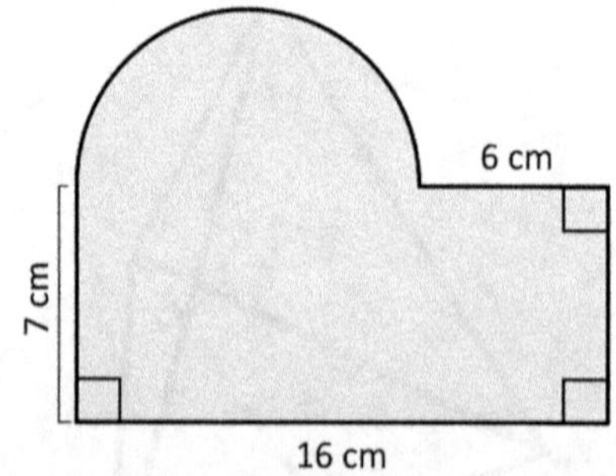

Area: ______________________

3.

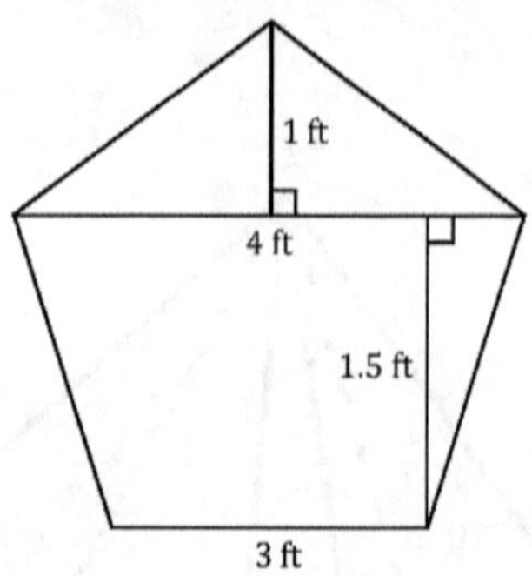

Area: ______________________

4.

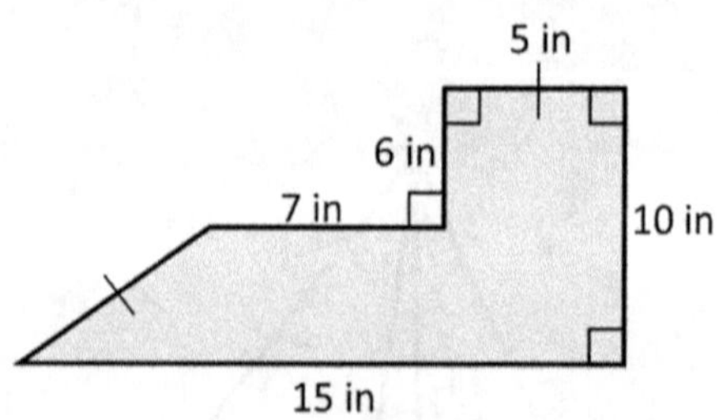

Area: ______________________

Chapter 5: Graphing & Data Analysis

Data refers to a collection of information. Data can be summarized using different graphical displays such as **histograms, pie charts, stem-and-leaf plots,** and **box-and-whisker plots.**

A **histogram** organizes data by showing the frequency of data in certain ranges. A histogram contains a vertical and horizontal axis. Typically, the frequency is listed vertically. Histograms can illustrate many types of data sets. It looks similar to a bar graph, but the bars are touching.

A **pie chart**, also known as a circle graph, is a type of graph that uses a circle to represent the total data and each portion of the circle is divided based on the way the data is distributed. Pie charts are commonly used with percentages and to illustrate how something is distributed across groups. A complete circle of the pie chart 100%, and the "slices" represent different portions.

A **stem-and-leaf plot** is a way to illustrate data with multi-digit values. This type of data representation is best used with a large number of data points that vary in value, say values from 0 to 100. In a stem-and-left plot, values are separated by place value. There is a list of numbers listed vertically along the left side; these are the stems. There is a list of numbers to the right of each stem; these are the leaves. There is a line dividing the stems from the leaves.

Box plots, or **box-and-whisker plots,** show the minimum value, the maximum value, the median value, the first quartile, and the third quartile of a data set. The minimum value is the smallest value and the maximum value is the largest value. The median is the exact center of a data set.

The **first quartile** is the median of the lower half of the data set, or the median of the values to the left of the median. The **third quartile** is the median of the upper half of the data set, or the median of the values to the right of the median.

The **measures of center** are different methods for estimating or measuring the center of a data set. The center of a data set can be described by the most frequent value in a data set, the value appearing in the middle of a data set arranged from least to greatest, or the value that is the average of that data set. The terms used to describe the center of a data set are mode, median, and mean.

The **mode** of a data set is the number that shows up most frequently.

Let's say we are given this data set.

$$\{0, 0, 1, 1, 1, 1, 1, 1, 1, 1, 1, 1, 2, 2, 2, 2, 3, 4, 5\}$$

The mode in a data set is the number that repeats the most frequently. If we look at this list of values, we can see that 1 shows up the most often, so the mode of this data set is 1.

The **median** of a data set is the literal middle value in a set of values. If you have a list of values, the median is going to be the value directly in the middle of a list of these values when the values are listed from least to greatest.

Let's say we have a set of data in a list.

$$\{3, 2, 2, 3, 0, 1, 2, 1, 1, 6, 2, 4, 2, 3, 6, 0, 0, 1, 1, 2, 4, 2, 4, 3, 1\}$$

This is a list of values, and we can determine the center, or median value in the data set by lining the values up from least to greatest and finding the middle value.

$$\{0, 0, 0, 1, 1, 1, 1, 1, 1, 2, 2, 2, 2, 2, 2, 2, 3, 3, 3, 3, 4, 4, 4, 6, 6\}$$

Now that we have our data set listed from least to greatest, the median will be the middle number in the set.

12 12

0 0 0 1 1 1 1 1 1 2 2 2 (2) 2 2 2 3 3 3 3 4 4 4 6 6

The center, or median value in this data set is 2, because when you list the values from least to greatest, 2 is the middle number. It is in the middle because it is the same distance from the smallest value to the greatest value and there are the same number of values, 12, on each side.

Finding the median of a data set that has even number of values requires a slightly different approach.

Let's say we are given this data set.

$$\{15, 16, 18, 20, 22, 12, 13, 14, 19, 21, 21, 11, 12, 14\}$$

First, we need to list the values from least to greatest to find the middle.

$$\{11, 12, 12, 13, 14, 14, 15, 16, 18, 19, 20, 21, 21, 22\}$$

Now, we can count to find the middle.

11 12 12 13 14 14 15|16 18 19 20 21 21 22

In this data set, there is not a middle number. Instead, the middle position lies between 15 and 16. So, to find the median value, we will have to find the mean of the two middle numbers, 15 and 16.

To find the mean of these middle values, we will add them together and divide by 2.

$$15 + 16 = 31$$
$$31 \div 2 = 15.5$$

So, the median of this data set is 15.5.

the data set together, and divide the sum by the total number of values.

In school, we are familiar with the mean of data sets because this is the way average grades are calculated. To get a final grade, often teachers will add the grade on each assignment or test together and divide by the total number of assignments or tests.

Let's say you are in a math class and your scores are listed in a data set.

$$\{76, 82, 82, 84, 87, 90, 85, 100, 94, 90\}$$

To find the mean, or average, of your scores, add the values and divide by the number of values. In this data set, if we add all the given grades together, we will get a total of 870. Now, if we divide 870 by the total number of grades, 10, we will get 87. That means that the final grade, the average grade, or the mean of the data set is 87.

The **spread** of a data distribution is a way to describe the range, or how spread out or close together the data is. The range of data is the difference between the smallest value and the largest value.

A **quartile** is one quarter of the values in a data set when they are arranged from least to greatest. Another way to describe the spread is by identifying the **interquartile range**, or IQR. The interquartile range is a way to describe the middle 50% of a data set. Thus, the interquartile range is the difference between the third quartile and the first quartile.

A **coordinate plane** is a way to graph data by plotting points on a grid. On a coordinate plane, there is an x-axis and a y-axis. The center point is called the **origin**, and its coordinates are (0,0). The x-axis runs horizontally, and the y-axis runs vertically. We use the x- and y-axis to graph ordered pairs. **Ordered pairs** are sets of values that tell us the coordinates of a point. They are written using parentheses and separated by a comma. (3,2) is an example of an ordered pair. In an ordered pair, the first number is the x-value, and the second number is the y-value. In this example, (3,2), $x = 3$ and $y = 2$. We use these values to plot points on a graph. The point (3,2) lies three units to the right of the origin along the x-axis and two units up along the y-axis. The point also lies at the intersection point of the lines $x = 3$ and $y = 2$. Positive values always run to the right along the x-axis and up along the y-axis. Negative values always run to the left along the x-axis and down along the y-axis.

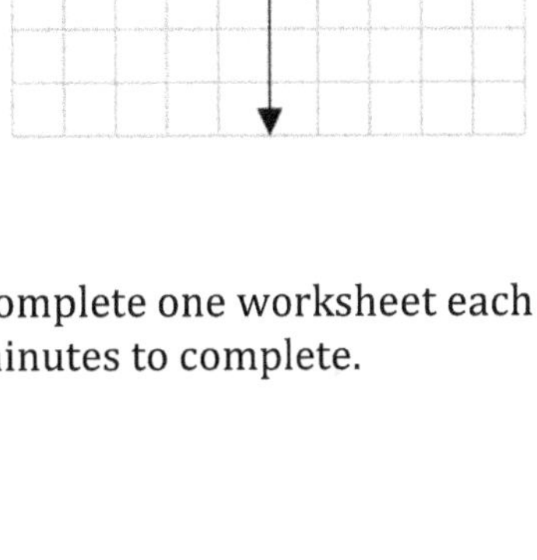

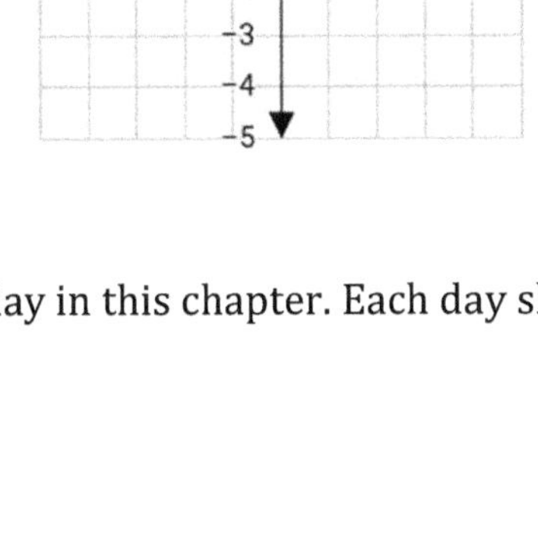

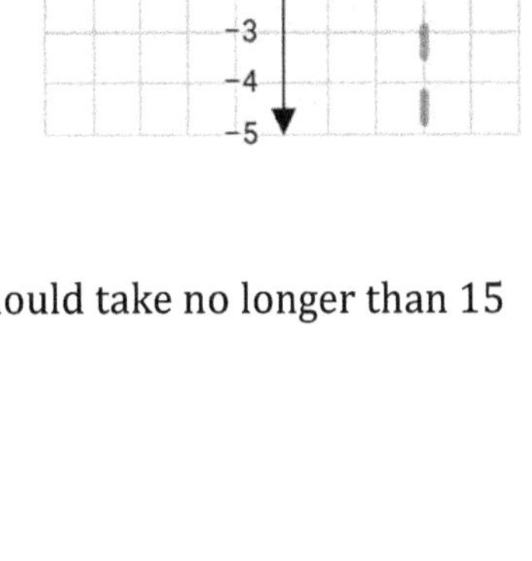

Complete one worksheet each day in this chapter. Each day should take no longer than 15 minutes to complete.

DAY 43

HISTOGRAMS

Given each data set, complete the following histograms.

1. Students recorded the amount of time they spent reading each day.

22, 35, 18, 50, 60, 42, 38, 35, 47, 55, 30, 33, 28, 45, 52, 36, 41, 27, 49

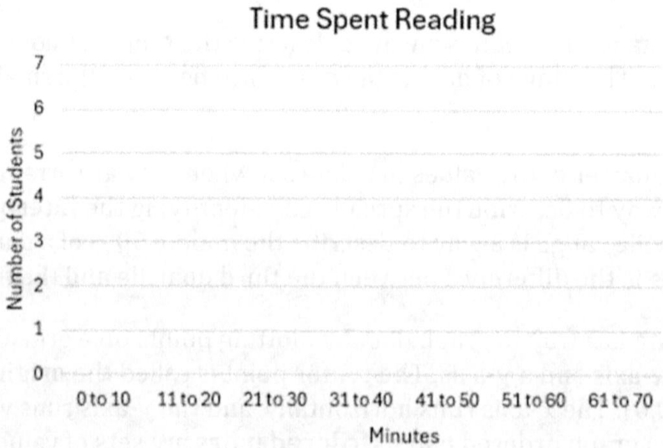

2. Students recorded the amount of time they spent playing video games each day.

15, 25, 35, 45, 55, 65, 20, 30, 66, 50, 60, 70, 18, 28, 36, 42, 0, 68, 22, 82

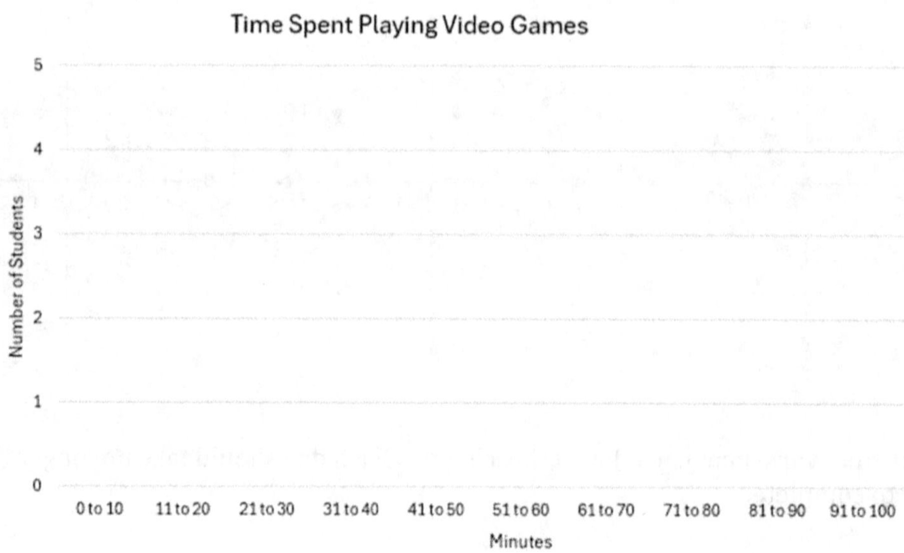

DAY 44

PIE CHARTS

Given each data set, complete the following pie charts.

1. Students were asked what their favorite school subject is. 8 students said math, 6 students said science, 5 students said English, 4 students said social studies, 3 students said art, and 4 students said P.E..

____

2. Students were asked how they get to school each morning. 9 students said they take the bus, 8 students said they take a car, 5 students said they walk, 3 students said they take a bike, and 2 students said they use some other form of transportation.

____

Day 45

Stem-and-Leaf Plots

Given each data set, complete the following stem-and-leaf plots.

1. A teacher was completing a measurement unit with her students. She had the students record their heights in inches.

65, 58, 65, 54, 55, 62, 56, 65, 67, 60, 61, 63, 64, 57

____

Stem	Leaf

Key: | =

2. A teacher was completing a financial literacy unit with her students. She had the students record the amount of money they get each week for completing household jobs.

$5, $17, $30, $27, $7, $18, $25, $8, $10, $22, $15, $20, $12

____

Stem	Leaf

Key: | =

Day 46

Box-and-Whisker Plots

Given each data set, complete the following box-and-whisker plots.

1. An Algebra 1 teacher gave a pop quiz, and the student scores are given below.

80, 90, 72, 74, 100, 77, 77, 78, 90, 92, 95, 98, 100, 80, 82, 88

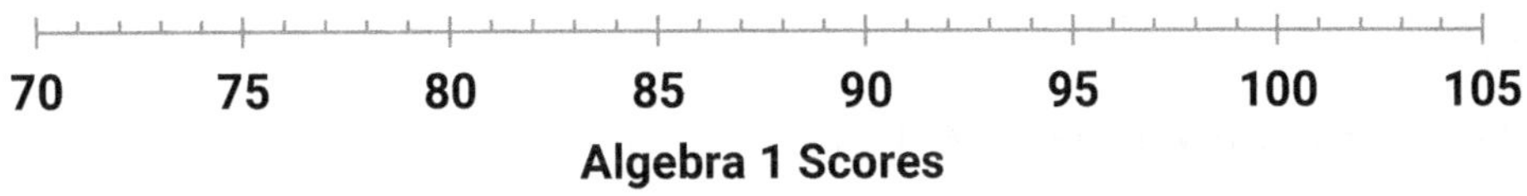

2. Students were asked how much time they spend doing homework each night. Below are the results.

60, 40, 40, 50, 50, 15, 22, 30, 30, 65, 70, 30, 55, 60, 60, 35, 40

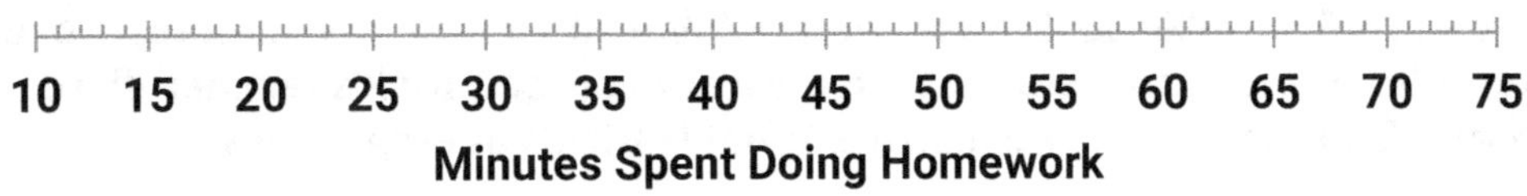

DAY 47

MEAN, MEDIAN, AND MODE

Answer the questions below.

1. Ten sample values were found to be 6, 7, 7, 9, 5, 7, 8, 2, 6, and 3. The mean and median values of these samples are:

a. 6 and 6.5
b. 6 and 7
c. 6 and 8
d. 6.5 and 7

2. What is the mode of the following numbers?

37, 46, 52, 52, 61, 63

a. 37
b. 52
c. 55
d. 311

3. Find the mean of the data set {76, 193, 83, 43, 105}.

a. 80
b. 83
c. 100
d. 500

4. Find the mode(s) of the data set {84, 29, 84, 93, 93, 84, 39, 84, 29, 93, 93, 29}.

a. 84
b. 93
c. 84 and 93
d. 29, 84, and 93

5. Identify the median for the data set {53, 81, 85, 82, 91, 72}.

a. 81.5
b. 81
c. 83.5
d. 77

6. Rick scores 95%, 68%, 86%, 83%, 64%, 92%, and 79% on his math tests over the semester. When calculating students' semester averages, Rick's teacher disregards each student's highest and lowest score. What is Rick's average test score?

a. 57.71%
b. 75.5%
c. 80.8%
d. 81.6%

Day 48

Range and Interquartile Range

Answer the questions below.

1. Consider the data set {196, 832, 736, 271, and 573}. Which of the following best describes the range of the data set?

a. 377
b. 636
c. 196
d. 832

2. Find and interpret the range for the data set {78, 29, 57, 10, 3, 84}.

a. The range of the data set is 6, which means that the values differ by a maximum of 6.
b. The range of the data set is 43.5, which means the average value in the set is 43.5.
c. The range of the data set is 81, which means the values of the set differ by no more than 81 units.
d. The range of the data set is 13.5, which means that the standard deviation is less than 13.5.

3. What is the interquartile range of the data below?

2, 4, 6, 8, 10, 12, 14, 16, 18, 20

a. 10
b. 11
c. 12
d. 13

4. What is the interquartile range for the following set of numbers?

2, 3, 4, 4, 5, 7, 8, 8, 11

a. 5
b. 4.5
c. 9
d. 5.8

5. What is the interquartile range of the data below?

1, 2, 4, 8, 11, 14, 14, 18, 19, 23

a. 8
b. 10
c. 12
d. 14

DAY 49

GRAPHING ON A COORDINATE PLANE

Place the following points on the graph below.

A(0, −4)	B(−4,0)	C(−1, −5)	D(6,3)	E(2, −6)
F(−5,4)	G(−2, −1)	H(−5, −5)	I(6, −3)	J(2, −2)

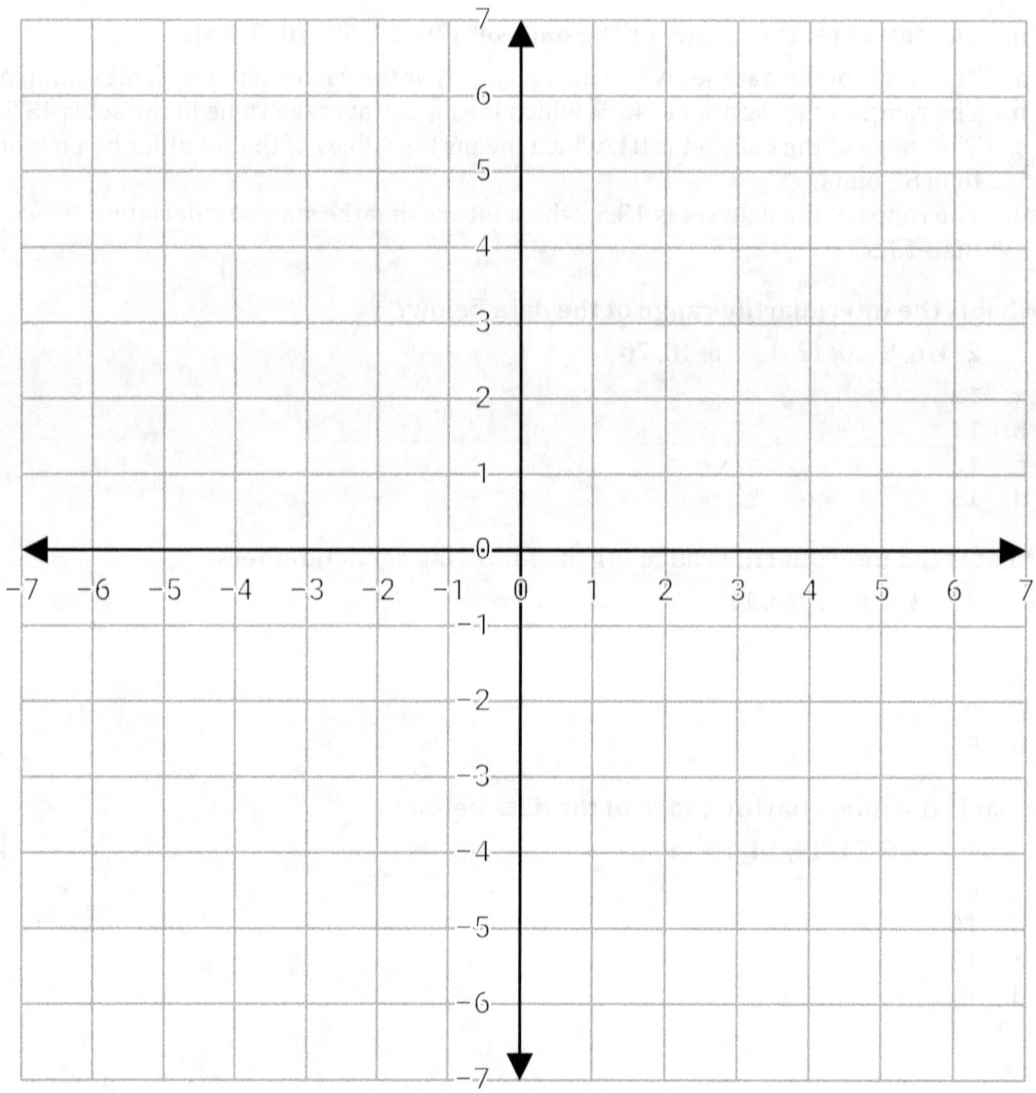

DAY 50

DETERMINING ORDERED PAIRS USING A COORDINATE PLANE

Determine the ordered pairs that are plotted on the coordinate plane below.

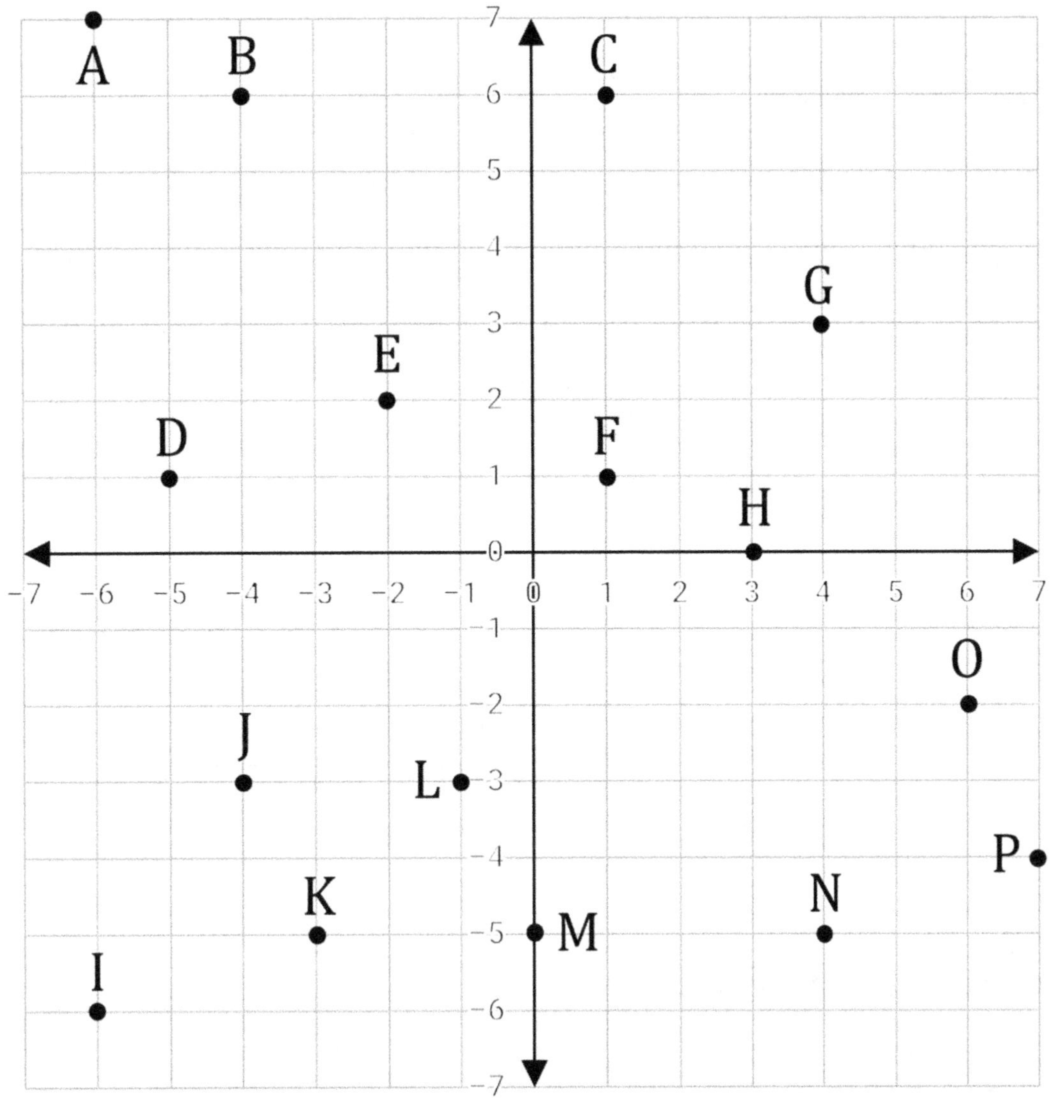

A ____________ B ____________ C ____________ D ____________

E ____________ F ____________ G ____________ H ____________

I ____________ J ____________ K ____________ L ____________

M ____________ N ____________ O ____________ P. ____________

Workbook Answers

Chapter 1: Numbers and Operations

Day 1

Classifying Numbers

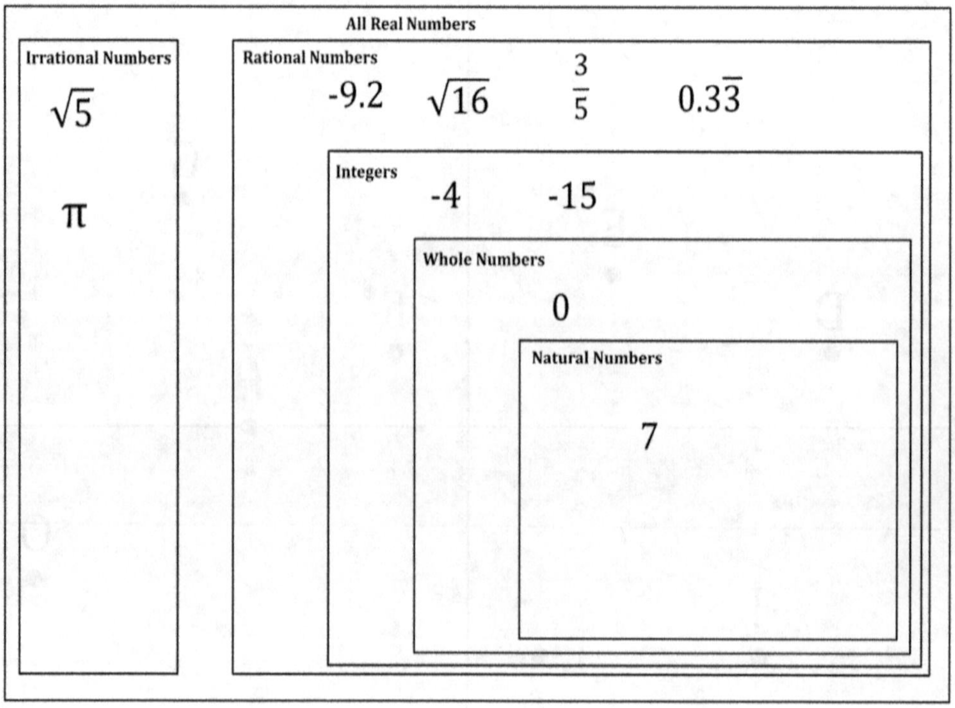

Day 2

Absolute Value and Opposites

1. 9
2. 12
3. 3.5
4. $\frac{2}{3}$
5. 0
6. -8
7. 4
8. 0
9. $\frac{4}{7}$
10. -5.6

Day 3

Exponents

1. 9
2. 125
3. 100
4. 16
5. 16
6. −8
7. −243
8. −16
9. 12
10. 32

Day 4

Square Roots

1. 7
2. 10
3. 3
4. 8
5. 6
6. 1
7. 9
8. 2
9. 4
10. 5

Day 5

Scientific Notation

1. 4.5×10^4
2. 6.7×10^6
3. 3.2×10^{-3}
4. 8.9×10^{-4}
5. 9.01×10^3
6. 32,000
7. 4,500,000
8. 0.0071
9. 0.09
10. 1,090

Day 6

Basic Operations

1. 3
2. -17
3. 11
4. -24
5. -6
6. $\frac{7}{8}$
7. $\frac{1}{2}$
8. $-\frac{6}{5}$
9. -3.6
10. 5

Day 7

Approximating Values

1. 4 and 5
2. 3 and 4
3. 7 and 8
4. 8 and 9
5. 9 and 10
6. 7.071
7. 3.162
8. 4.472
9. 12.247
10. 8.124

Day 8

Ordering Real Numbers

1. $-8, -3, 0, 2, 5$
2. $4.002, 4.02, 4.2, 4.22$
3. $-\frac{3}{4}, -0.6, -\frac{1}{2}, -0.25$
4. $2.9, \sqrt{9}, 3.1, \pi$
5. $-4.1, -\sqrt{16}, -3$

Day 9

Comparing Real Numbers

1. >
2. <
3. =
4. <
5. >
6. =
7. >
8. <
9. >
10. <
11. <
12. <

Day 10

Order of Operations

1. 3
2. 31
3. 2
4. 5
5. 40
6. 25
7. 9
8. 2

Chapter 2: Expressions, Equations, and Inequalities

Day 11

Writing Expressions

1. $s + 8$
2. $6n$
3. $m - 12$
4. $5h$
5. $3t + 2$
6. $10t + 3$
7. $15m + 25$
8. $x^2 + 4$

Day 12

Simplifying Expressions

1. $8x + 5y$
2. $2a + 6$
3. $37m - 3$
4. $7x + 2$
5. $10b + 4$
6. $4n + 13$
7. $9x + 15$
8. $8p - 1$
9. $10x + 2y$
10. $8x + 12$

Day 13

Writing One-Variable Equations

1. $25m + 30 = 155$
2. $m + 18 = 52$
3. $35h + 120 = 245$
4. $15h = 105$
5. $p - 7 = 15$
6. $6n + 12 = 54$
7. $6h + 1 = 13$
8. $b - 9 + 21$

Day 14

Solving One-Variable Equations

1. $x = 5$
2. $x = 14$
3. $x = 6$
4. $x = 30$
5. $x = 5$
6. $x = 6$
7. $x = 4$
8. $x = 3$
9. $x = 12$
10. $x = 6$

Day 15

Writing One-Variable Inequalities

1. $x < 12$
2. $f \geq 3$
3. $4d \geq 28$
4. $4n + 2 < 30$
5. $p \leq 50$
6. $s > 500$
7. $3p > 60$
8. $480 \div t \leq 60$

Day 16

Solving One-Variable Inequalities

1. $x > 7$
2. $x \leq 13$
3. $x < 6$
4. $x \geq 14$
5. $x > 4$
6. $x \leq 5$
7. $x \geq 3$
8. $x < 3$
9. $x < -4$
10. $x \geq -32$

Day 17

Graphing One-Variable Inequalities

1. $x \leq 4$

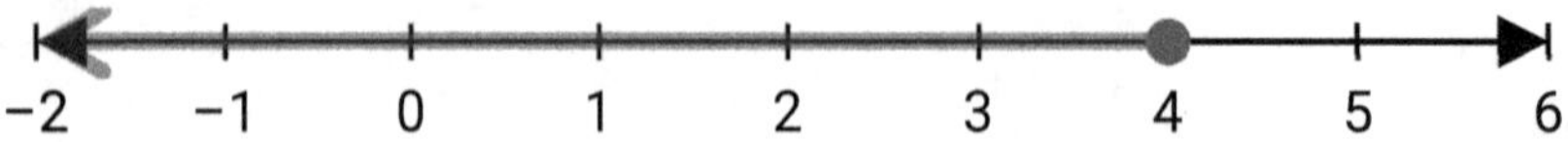

2. $x > 9$

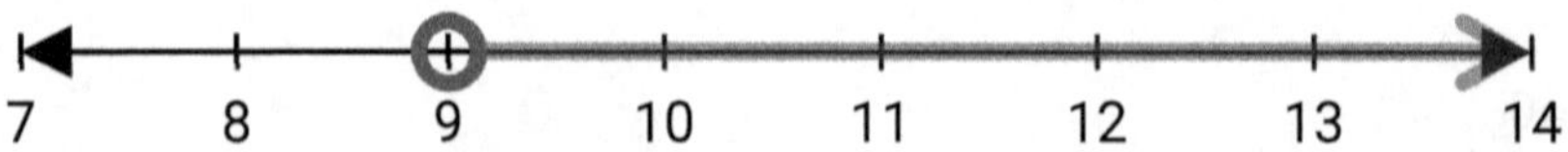

3. $x < -8$

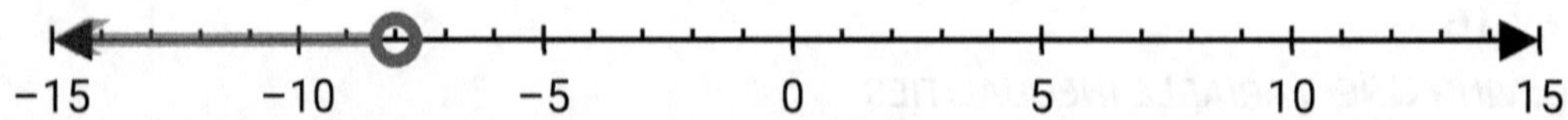

4. $x \geq -8$

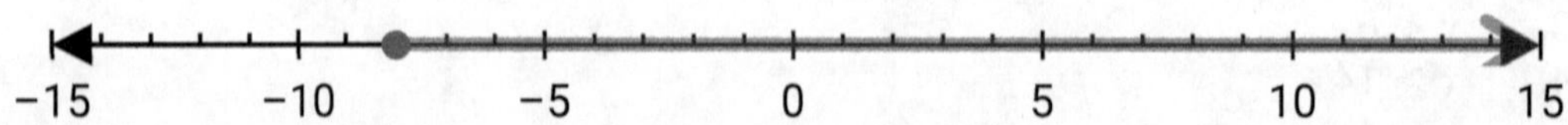

5. $x \leq -8$

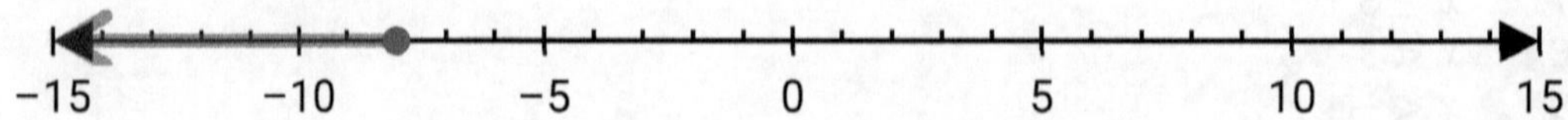

Day 18

Constant of Proportionality

1. 5
2. 7
3. 2.5
4. 3
5. 4
6. 1.5
7. $\frac{1}{2}$
8. -5
9. 7
10. $\frac{1}{2}$
11. -3
12. 5

Day 19

Proportional or Non-Proportional

1. Proportional
2. Non-proportional
3. Proportional
4. Proportional
5. Non-proportional
6. Proportional

Day 20

Simplifying Two-Variable Equations

1. $y = -\frac{5}{4}x + 5$
2. $y = -\frac{3}{2}x + 15$
3. $y = 8x - 18$
4. $y = 3x - 6$
5. $y = -\frac{8}{3}x + 8$
6. $y = -\frac{1}{5}x + 2$
7. $y = -\frac{11}{6}x + 9$
8. $y = -\frac{9}{2}x + 7$
9. $y = -\frac{1}{2}x + 3$
10. $y = \frac{11}{4}x - 5$

Day 21

Graphing Two-Variable Equations

1.

x	y
−2	−4
−1	−2
0	0
1	2

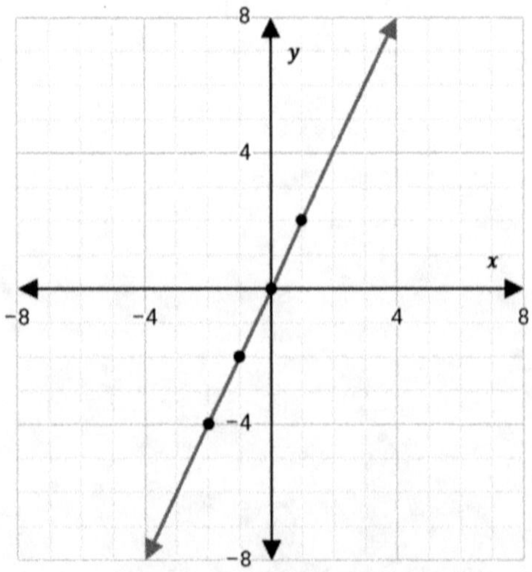

3.

x	y
−2	1
−1	2
0	3
1	4

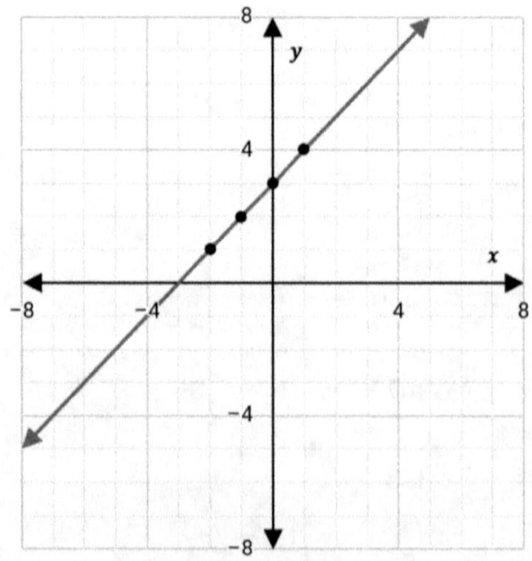

2.

x	y
−2	4
−1	1
0	−2
1	−5

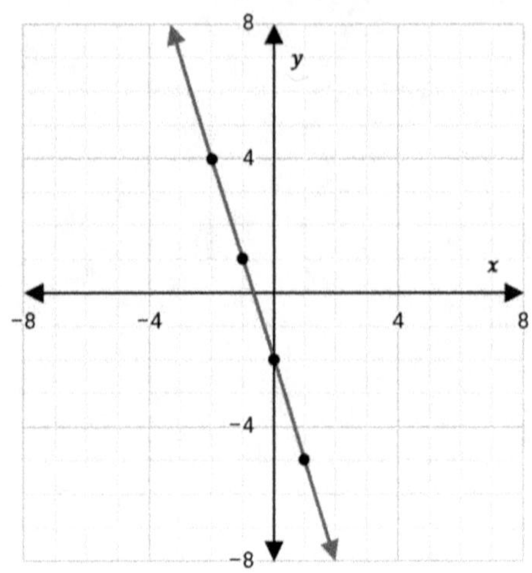

4.

x	y
−2	−1
−1	$-\frac{1}{2}$
0	0
1	$\frac{1}{2}$

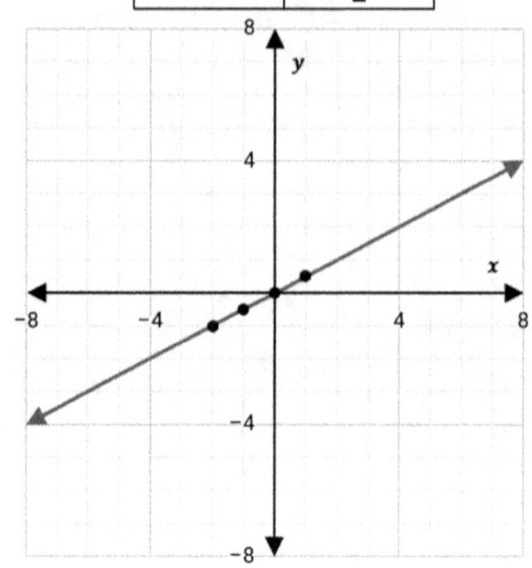

Day 22

Substitution

1. $y = 6$
2. $x = 4$
3. $y = 5$
4. $x = 1$
5. $y = 3$
6. $x = 6$
7. $y = 5$
8. $x = 3$

Chapter 3: Proportional Reasoning

Day 23

Ratios and Rates

1. $1:4$
2. $3:2$
3. $4:7$
4. 60 miles per hour
5. $2 per notebook
6. 12 students per class
7. 5 miles per hour
8. $0.50 per apple
9. 3 gallons per minute
10. 120 pages

Day 24

Proportions

1. $x = 2$
2. $x = 15$
3. $x = 21$
4. $x = 2$
5. $x = 30$
6. $x = 21$
7. $x = 20$
8. $x = 10$
9. $x = 10$
10. $4.50

DAY 25

PERCENT PROPORTIONS – FINDING THE PART

1. 20
2. 30
3. 9
4. 36
5. 24
6. 6
7. $13.80
8. $3,440
9. $18
10. $21.60

DAY 26

PERCENT PROPORTIONS – FINDING THE WHOLE

1. 80
2. 50
3. 60
4. 120
5. 30
6. 120
7. $54,118
8. 30
9. $2,667
10. 6

DAY 27

PERCENT PROPORTIONS – FINDING THE PERCENT

1. 20%
2. 30%
3. 25%
4. 40%
5. 75%
6. 20%
7. 70%
8. 60%
9. 19.4%
10. 41.7%

DAY 28

SIMILAR FIGURES

1. $x = 36$ ft
2. $x = 15$ units
3. $\overline{AC} = 12$ cm

Day 29

Scale Drawings

1. 40 ft
2. 392 in^2
3. 16 ft

Day 30

Dilation on a Coordinate Plane

1. 2
2. $\frac{1}{2}$

DAY 31

SAMPLE SPACES – LISTS, TABLES, AND TREE DIAGRAMS

1. 36 possible combinations

1, 1	1, 2	1, 3	1, 4	1, 5	1, 6
2, 1	2, 2	2, 3	2, 4	2, 5	2, 6
3, 1	3, 2	3, 3	3, 4	3, 5	3, 6
4, 1	4, 2	4, 3	4, 4	4, 5	4, 6
5, 1	5, 2	5, 3	5, 4	5, 5	5, 6
6, 1	6, 2	6, 3	6, 4	6, 5	6, 6

2. 9 possible combinations

	Hot Dog (HD)	Hamburger (H)	BBQ (B)
Chips (C)	C, HD	C, H	C, B
Potato Salad (PS)	PS, HD	PS, H	PS, B
Side Salad (SS)	SS, HD	SS, H	SS, B

3. 8 possible combinations

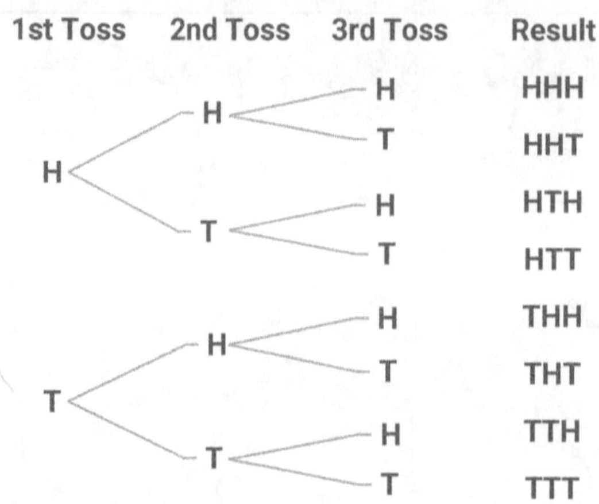

Day 32

Simple Probability

1. $\frac{1}{6}$
2. $\frac{1}{2}$
3. $\frac{1}{2}$
4. $\frac{3}{5}$
5. $\frac{2}{5}$
6. $\frac{1}{2}$
7. $\frac{1}{2}$
8. $\frac{1}{3}$
9. $\frac{1}{5}$
10. $\frac{1}{2}$

Chapter 4: Geometry

Day 33

Identifying Parts of a Circle

1.	Circle: A	Radius: AB, AC, AD	Diameter: BC
2.	Circle: E	Radius: EF, EH, EG	Diameter: FG
3.	Circle: X	Radius: XW, XZ, XY	Diameter: WY
4.	Circle: R	Radius: RQ, RS,RT	Diameter: QT
5.	Circle: M	Radius: ML, MN, MO	Diameter: LO
6.	Circle: X	Radius: XR, XA, XP	Diameter: AR

Day 34

Finding the Radius of a Circle

1. $18 \div 2 = 9$ cm
2. $36 \div 2 = 18$ in
3. $112 \div 2 = 56$ ft
4. $388 \div 2 = 194$ in
5. $956 \div 2 = 478$ ft
6. $3{,}624 \div 2 = 1{,}812$ cm

Day 35

Finding the Diameter of a Circle

1. $12 \times 2 = 24$ in
2. $29 \times 2 = 58$ cm
3. $89 \times 2 = 178$ ft
4. $113 \times 2 = 226$ cm
5. $624 \times 2 = 1{,}248$ in
6. $2{,}935 \times 2 = 5{,}870$ ft

Day 36

Finding the Circumference of a Circle

1. 345.4 in
2. 226.08 cm
3. 942 ft
4. 120.89 ft
5. 3,337.192 in
6. 21.64716 cm

Day 37

Area of a Circle

1. $A = \pi(3)^2 = 28.27 \text{ in}^2$
2. $A = \pi(5)^2 = 78.54 \text{ in}^2$
3. $A = \pi(12)^2 = 452.39 \text{ cm}^2$
4. $A = \pi(4)^2 = 50.27 \text{ in}^2$
5. $A = \pi(20)^2 = 1{,}256.64 \text{ cm}^2$
6. $A = \pi(6)^2 = 113.10 \text{ in}^2$

Day 38

Surface Area and Volume of Rectangular Prisms

1. Surface Area: 542 ft^2
   Volume: 819 ft^3
2. Surface Area: 265.34 m^2
   Volume: 292.07 m^3
3. Surface Area: 142 cm^2
   Volume: 105 cm^3
4. Surface Area: 972 m^2
   Volume: 1,417.5 m^3

Day 39

Volume of Triangular Prisms

1. 40 m^3
2. 336 m^3
3. 1,102.5 cm^3
4. 1,500 cm^3

Day 40

Surface Area of Triangular Prisms

1. 336 cm^2
2. 330 in^2
3. 372 ft^2
4. 208 m^2

Day 41

Surface Area of Pyramids

1. 60 cm^2
2. 112 in^2
3. 360 ft^2
4. 288 m^2

Day 42

Area of Composite Figures

1. 136 in^2
2. 131.24 cm^2
3. 13.25 ft^2
4. 158 in^2

Chapter 5: Graphing & Data Analysis

DAY 43

HISTOGRAMS

1.

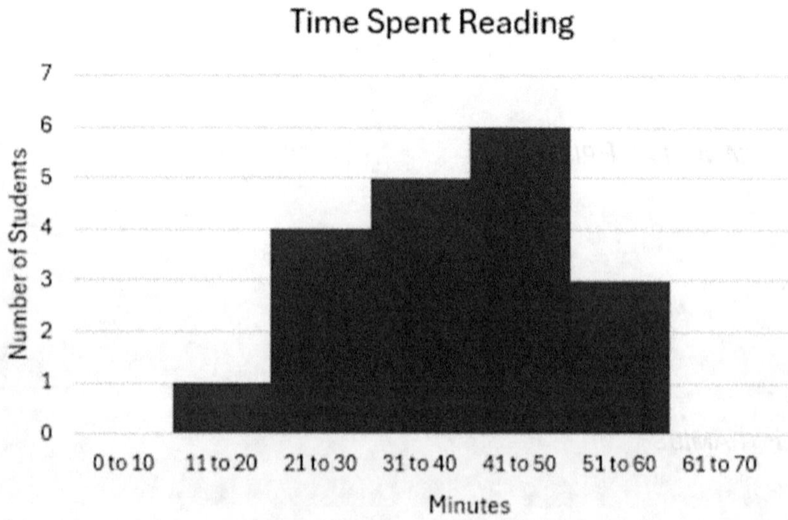

2.

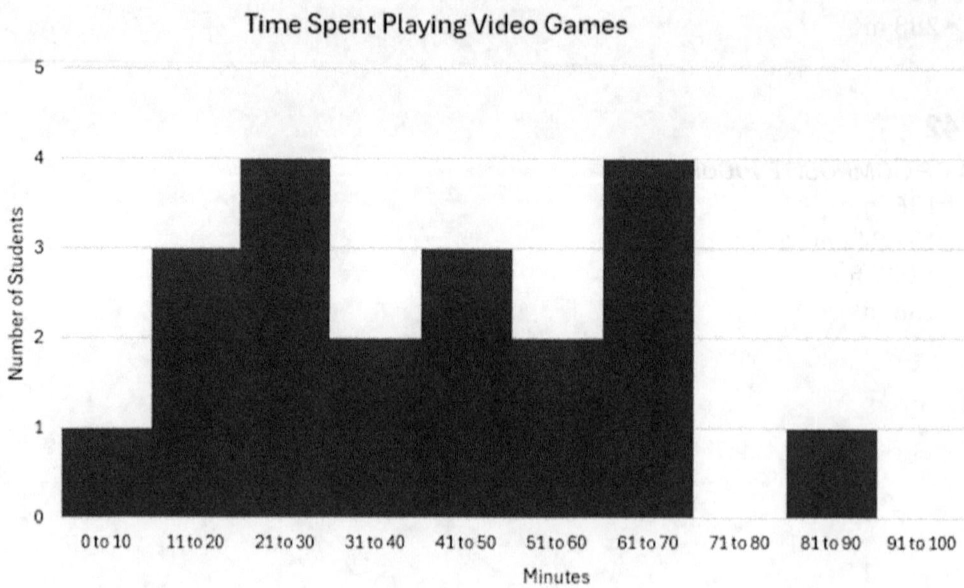

Day 44

Pie Charts

1.

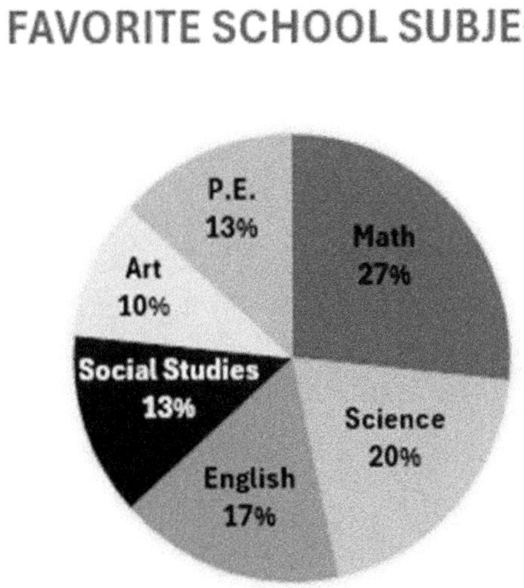

2.

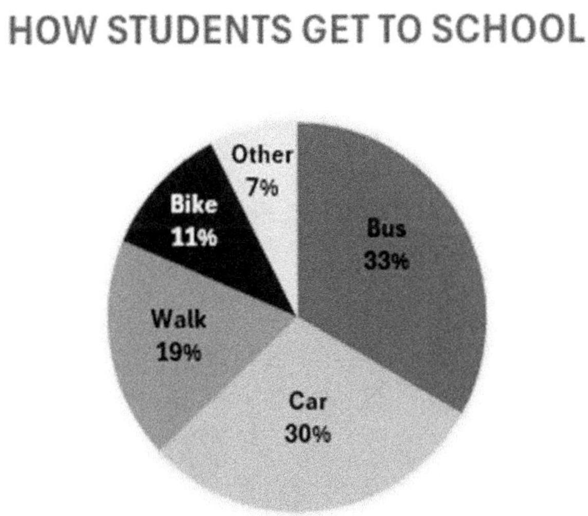

DAY 45

STEM-AND-LEAF PLOTS

1.

Student Height

Stem	Leaf
5	4 5 6 7 8
6	0 1 2 3 4 5 5 5 7

Key: 5 | 4 = 54 inches

2.

Student Chore Earnings

Stem	Leaf
0	5 7 8
1	0 2 5 7 8
2	0 2 5 7
3	0

Key: 3 | 0 = $30

Day 46

Box-and-Whisker Plots

1.

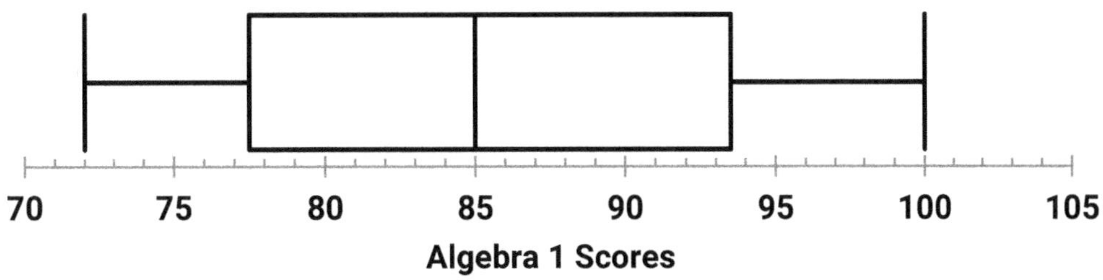

2.

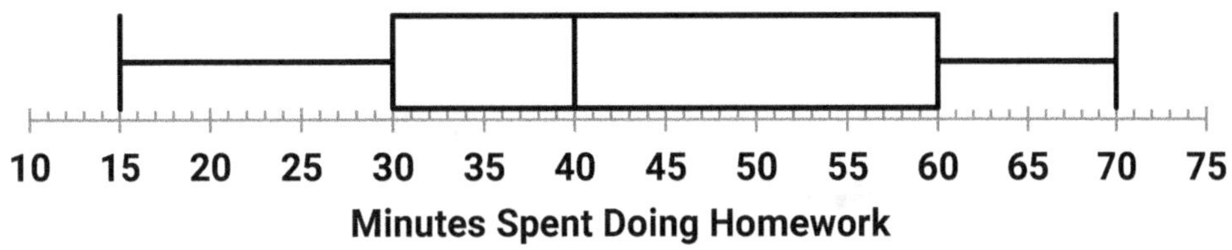

Day 47

Mean, Median, and Mode

1. A: 6 and 6.5
2. B: 52
3. C:100
4. C: 84 and 93
5. A: 81.5
6. D: 81.6%

Day 48

Range and Interquartile Range

1. B: 636
2. C: The range of the data set is 81, which means the values of the set differ by no more than 81 units.
3. A: 10
4. B: 4.5
5. D: 14

Day 49

Graphing on a Coordinate Plane

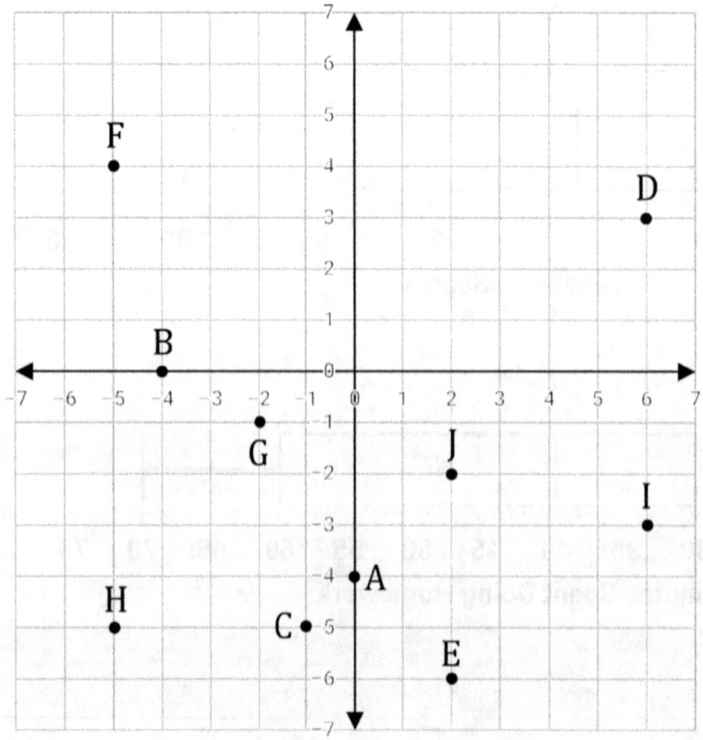

Day 50

Determining Ordered Pairs Using a Coordinate Plane

A. (−6,7)
B. (−4,6)
C. (1,6)
D. (−5,1)
E. (−2,2)
F. (1,1)
G. (4,3)
H. (3,0)
I. (−6,−6)
J. (−4,−3)
K. (−3,−5)
L. (−1,−3)
M. (0,−5)
N. (4,−5)
O. (6,−2)
P. (7,−4)

www.ingramcontent.com/pod-product-compliance
Lightning Source LLC
LaVergne TN
LVHW082357200926
848481LV00012B/3591